Long COVID Fatigue

Thorsten Rudroff

Long COVID Fatigue

Clinical Sciences, Artificial Intelligence and the Future of Brain Health

Springer

Thorsten Rudroff
Turku PET Centre
Turku University Hospital
Turku, Finland

ISBN 978-3-031-89473-2 ISBN 978-3-031-89474-9 (eBook)
https://doi.org/10.1007/978-3-031-89474-9

This Springer imprint is published by the registered company Springer Nature Switzerland AG
The registered company address is: Gewerbestrasse 11, 6330 Cham, Switzerland

My deepest gratitude goes to my wife, Nona, whose unconditional love and support have made this work possible. Her trust, patience, and understanding have sustained me through challenging times and celebrated with me during moments of joy. Thank you, Nona, for being my anchor and my inspiration—this achievement is as much yours as it is mine.

Preface

The emergence of Long COVID has presented the medical and research communities with one of the most significant health challenges of our time. This book represents a comprehensive examination of Long COVID fatigue, bringing together current research, clinical observations, and emerging therapeutic approaches to address this debilitating condition that affects millions worldwide.

As a researcher who has dedicated significant time to studying the neurological impacts of Long COVID, I have witnessed firsthand how this condition challenges our traditional understanding of post-viral syndromes and forces us to reconsider our approaches to fatigue assessment and treatment. The complexity of Long COVID fatigue—its varied presentations, multiple contributing factors, and impact across different patient populations—demands a thorough and nuanced exploration.

This book aims to bridge the gap between laboratory findings and clinical practice, offering healthcare providers, researchers, and students a detailed understanding of Long COVID fatigue's mechanisms, manifestations, and management strategies. Each chapter builds upon the previous ones to create a coherent narrative, from basic pathophysiology through cutting-edge therapeutic approaches.

A particular emphasis has been placed on emerging technologies and methodologies, especially the role of artificial intelligence in advancing our understanding and treatment of Long COVID fatigue. The integration of AI with traditional clinical approaches represents a promising frontier in managing this challenging condition.

The text also addresses crucial aspects often overlooked in medical literature, including sex-based differences in Long COVID fatigue manifestation, age-related considerations, and the complex interplay between physical and cognitive symptoms. Special attention has been given to practical clinical applications, ensuring that theoretical insights translate into actionable treatment strategies.

Throughout this work, I have attempted to maintain a balance between academic rigor and clinical utility, supporting assertions with current research while acknowledging areas where our understanding remains incomplete. The rapidly evolving nature of Long COVID research means that some findings presented here may be refined or revised as new evidence emerges.

It is my hope that this book will serve not only as a current reference but also as a foundation for future research and clinical innovation in the field of Long COVID fatigue. As we continue to face the challenges of this condition, our success will depend on our ability to integrate new findings, adapt our approaches, and maintain our commitment to evidence-based, patient-centered care.

Turku, Finland
February 2025

Thorsten Rudroff

Acknowledgments

I would like to express my sincere gratitude to everyone who has contributed to the realization of this book. In particular, I wish to thank Andrea Ridolfi and Pinky Sathishkumar from Springer, part of Springer Nature, for their unwavering support, expert guidance, and patience throughout the development of this book project. Their editorial expertise and commitment to excellence have been invaluable in bringing this work to fruition.

Special thanks go to my colleagues from the Turku PET Centre, University of Turku, Turku University Hospital, Finland—Drs. Riku Klén, Oona Rainio, and Jetro Tuulari. Our stimulating discussions about artificial intelligence and neuroimaging techniques have significantly enriched this book's content and helped shape its direction. Their insights and expertise in advanced neuroimaging methodologies have been particularly valuable in developing the chapters on AI applications and brain imaging analysis.

This book has benefited immensely from the contributions of numerous clinicians, researchers, and healthcare professionals who have shared their experiences and insights. Their work on the frontlines of Long COVID care has been instrumental in advancing our understanding of this challenging condition.

I would also like to acknowledge the patients who have participated in Long COVID research studies. Their willingness to share their experiences and participate in clinical trials has been crucial in advancing our knowledge of this condition. Their stories and struggles have reminded us of the human impact of Long COVID and the importance of our work in this field.

Finally, I want to express my deepest gratitude to my family. To my father, Dieter Rudroff, whose lifelong dedication to hard work has been a constant source of inspiration. To the memory of my mother Eike Rudroff, who instilled in me a passion for helping others and whose influence continues to guide my work. To my son, Tyler Rudroff, whose curiosity and enthusiasm remind me daily of the importance of clear scientific communication and the impact our research can have on future generations.

Contents

About the Author

Thorsten Rudroff is a Professor (Docent) in Experimental Neuroimaging at the Turku PET Centre, University of Turku and Turku University Hospital, Finland. He previously held positions as Associate Professor at the University of Iowa and Assistant Professor at Colorado State University, USA. With over 20 years of experience in neurophysiology research, he has made significant contributions to understanding fatigue and cognitive dysfunction in neurological disorders, particularly in Multiple Sclerosis and Long COVID. His scientific work includes over 85 publications in prestigious journals. Dr. Rudroff received his Ph.D. in Neurophysiology from the University of Constance, Germany, and completed his postdoctoral training at the University of Colorado, Boulder. He is particularly passionate about leveraging artificial intelligence to advance medical research and improve patient outcomes. Dr. Rudroff serves on multiple editorial boards and has received several significant research grants from the National Institutes of Health. His current research focuses on the intersection of neuroimaging, artificial intelligence, and Long COVID, with particular emphasis on developing innovative approaches to understand and treat fatigue symptoms. Furthermore, his focus is on developing AI-driven approaches for early detection of Alzheimer's disease through neuroimaging biomarkers.

List of Figures

List of Tables

"A Mysterious Malaise" 1

1.1 Long COVID

On March 11, it will have been 4 years since the WHO officially declared COVID-19 as a pandemic. Despite global efforts to move beyond the initial acute phase, an estimated 65 million individuals continue to grapple with the effects of "Long COVID." The World Health Organization (WHO) defines Long COVID as the continuation or development of new symptoms 3 months after the initial SARS-CoV-2 infection, with these symptoms lasting for at least 2 months with no other explanation [1]. This condition, which often manifests with symptoms such as persistent fatigue, shortness of breath, and cognitive impairment, can severely hinder a person's ability to carry out daily activities for months or even years. While the majority of individuals infected with SARS-CoV-2 typically recover within a few weeks, long COVID is estimated to afflict 10–20% of cases, affecting individuals of all age groups, including children. It is noteworthy that many long COVID cases arise from initially mild acute illness [1].

1.2 The Societal and Economic Impact of Long COVID

Long COVID has emerged as one of the most significant public health challenges of our time, with far-reaching implications that extend well beyond individual health outcomes. Understanding its true impact has been challenging, as few countries have established comprehensive surveillance systems to estimate the burden of Long COVID at the population level. Most available data focus on economic and healthcare system impacts rather than precise case numbers. Nevertheless, the evidence we do have paints a concerning picture.

With an estimated 400 million cases globally by the end of 2023, representing approximately 6–7% of adults and 1% of children affected, the condition has created ripple effects throughout society and the global economy. Two key

T. Rudroff, *Long COVID Fatigue*, https://doi.org/10.1007/978-3-031-89474-9_1

visualizations help illustrate the scale and nature of this impact. Figure 1.1 traces the global accumulation of Long COVID cases over time, while Fig. 1.2 compares its economic toll across major Western economies.

Figure 1.1 shows the dramatic increase in global Long COVID cases from 2020 to 2023, with projections for 2024. The visualization illustrates several key points from Al-Aly [2]:

1. The rapid acceleration of cases, particularly between 2020 and 2021
2. The continued but somewhat slower growth in 2022 and 2023
3. The substantial cumulative burden, reaching 409 million cases by 2023

The societal impact has been profound. Long COVID has strained healthcare systems worldwide, leading to increased wait times and delays in essential care. The condition disproportionately affects marginalized communities, exacerbating existing health inequities. It has also disrupted family dynamics, as many individuals struggle with daily activities and require support from family members who become informal caregivers.

Furthermore, Long COVID threatens progress toward the United Nations' Sustainable Development Goals (SDGs) by impacting poverty reduction, healthcare access, education, gender equality, and economic growth. The condition has particularly affected women, who are more likely to experience certain forms of Long COVID and face greater setbacks in career advancement and economic security.

The economic toll is staggering. Conservative estimates suggest that Long COVID costs the global economy approximately $1 trillion annually, equivalent to roughly 1% of global GDP. This impact stems from multiple sources: reduced workforce participation, with studies showing that people with Long COVID are 10% less likely to be employed and work 25–50% fewer hours when employed; substantial healthcare costs; and increased disability claims. In the United States alone, between 2 and 4 million people were estimated to be out of work due to Long COVID in 2022, while in the United Kingdom, nearly 20% of affected individuals were unable to work at all.

Figure 1.2 presents a striking comparison of Long COVID's economic impact across five major economies—the United States and four European nations. The most notable feature is the dramatic difference in work hours lost between the

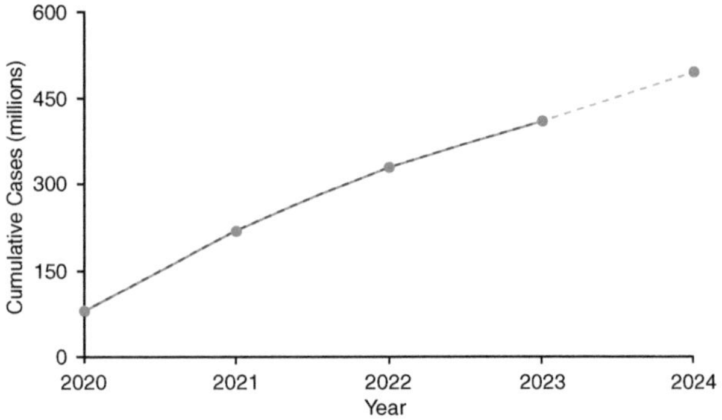

Fig. 1.1 Global cumulative incidence of Long COVID. (Data Source: Global Burden of Disease Long COVID Collaborators, Nature Medicine [2] 2024 Projection Range: 450–580 million cases)

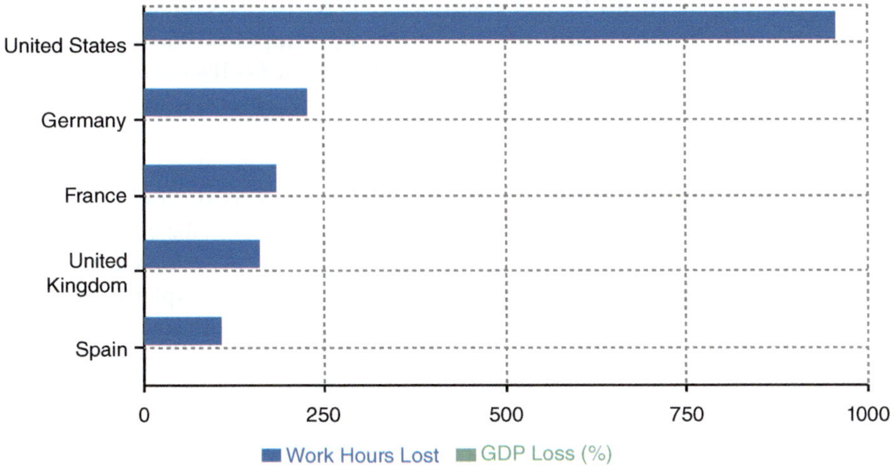

Fig. 1.2 Long COVID impact: Europe and USA (2024). (Data Source: Nature Medicine [2]—Economic Impact Analysis)

United States (953.6 million hours) and European countries, with Germany (225.4 million), France (182.8 million), the United Kingdom (158.9 million), and Spain (106.0 million) showing significantly lower figures. This substantial difference likely reflects both the larger U.S. population and different approaches to measuring Long COVID's workplace impact. However, when examining GDP loss percentages, Germany leads with 0.7%, followed by France at 0.6%, while the United States, United Kingdom, and Spain all show 0.5% GDP loss, suggesting similar relative economic burdens across these developed economies.

The combined evidence from both global trends and country-specific analyses reveals Long COVID as a complex crisis that operates on multiple levels—from individual health impacts to macroeconomic disruption. While the economic costs can be quantified in terms of lost work hours and GDP impact, the societal toll—including strained healthcare systems, disrupted families, and exacerbated inequalities—may have even longer-lasting implications. As countries grapple with these challenges, the data suggests that effective responses will require both national commitment and international cooperation, particularly given the shared nature of the burden across developed economies. The varying impacts across different metrics and contexts emphasize the need for nuanced, adaptable approaches to both measuring and addressing Long COVID's ongoing effects on society and economy.

1.3 Long COVID Fatigue

Among the myriad symptoms associated with Long COVID, fatigue stands out as a particularly debilitating and prevalent manifestation. Fatigue in Long COVID is characterized by a profound, persistent exhaustion that does not improve with rest and can significantly impair an individual's ability to perform daily activities [1].

Sudre et al. [3] found that 13.3% of confirmed COVID-19 cases reported symptoms lasting beyond 28 days, with 4.5% experiencing symptoms persisting beyond 8 weeks. Notably, Huang et al. [4] observed that 76% of COVID-19 survivors experienced at least one lingering symptom 6 months after recovery, with fatigue being among the most common complaints.

Despite global efforts to move beyond the initial acute phase, an estimated 65 million individuals continue to grapple with the effects of "Long COVID." The World Health Organization (WHO) defines long COVID as the continuation or development of new symptoms 3 months after the initial SARS-CoV-2 infection, with these symptoms lasting for at least 2 months with no other explanation [1].

The emergence of COVID-19 and subsequent recognition of Long COVID represent one of the most rapidly evolving medical challenges in recent history. The story begins in December 2019, when the first cases of a novel respiratory illness were reported in Wuhan, China. As documented by Wu et al. [5] in Nature, this new coronavirus would soon reshape global health perspectives and healthcare systems worldwide.

The gravity of the situation became apparent in early 2020, leading the World Health Organization to declare a Public Health Emergency of International Concern in January. By March 11, 2020, as the virus spread rapidly across continents, WHO Director-General Dr. Tedros Adhanom Ghebreyesus made the historic announcement declaring COVID-19 a global pandemic.

A crucial development emerged in May 2020, when patient groups and clinicians began reporting persistent symptoms lasting weeks or months after the initial infection. This grassroots recognition of what would become known as Long COVID demonstrated the essential role of patient advocacy in advancing medical understanding. As Perego et al. [6] noted in their opinion piece, the term "Long Covid" itself originated from patient communities, highlighting the collaborative nature of this medical discovery.

The scientific community responded swiftly to these emerging reports. August 2020 saw the publication of the first major research on persistent symptoms, with Carfì and colleagues [7] documenting their findings in JAMA. This was followed by the establishment of specialized Long COVID clinics in October 2020, as healthcare systems began adapting to address this new chronic condition.

The year 2021 marked several crucial developments in Long COVID recognition and research. In January, the WHO officially acknowledged Long COVID as a condition requiring specific attention and study. By April, landmark cohort studies, such as the work by Al-Aly et al. [8] published in Nature, began providing comprehensive characterizations of post-acute COVID-19 sequelae.

A significant milestone occurred in October 2021, when the WHO established a clinical case definition for post-COVID-19 condition through a global Delphi consensus. This standardization of terminology and diagnostic criteria proved crucial for advancing research and treatment approaches. The year concluded with the launch of major global research initiatives, including the NIH's comprehensive study of Long COVID.

The period from 2022 through 2024 has seen rapid advancement in understanding and treating Long COVID. The RECOVERY Collaborative Group's 2022 treatment trials marked the beginning of systematic therapeutic investigations. By 2023, sophisticated biomarker studies, such as those published by Su et al. [9], were providing crucial insights into the biological mechanisms underlying Long COVID.

Moving into 2024, new therapeutic approaches continue to emerge, building on the accumulated knowledge of the previous years. This ongoing evolution of treatment strategies reflects the medical community's growing understanding of Long COVID's complex pathophysiology and the need for personalized therapeutic approaches.

Throughout this timeline, we see the remarkable speed at which medical knowledge and clinical practice have evolved in response to this novel health challenge. From initial recognition to systematic study and treatment development, the Long COVID timeline demonstrates both the capabilities and adaptability of modern medical science, while also highlighting the crucial role of patient advocacy in advancing medical understanding. This historical progression continues to inform current approaches to diagnosis, treatment, and research into post-viral conditions.

Figure 1.3 shows a chronological timeline depicting major developments in the understanding and recognition of COVID-19 and Long COVID. The central horizontal line represents time progression from December 2019 to 2024. Red circles indicate major milestone events. Alternating text placement (above and below the timeline) shows key developments including initial disease identification (December 2019), WHO declarations (2020), the emergence of Long COVID as a recognized condition (2020), the establishment of clinical definitions (2021), and research advances (2022–2024). This progression illustrates the rapid evolution from the initial disease outbreak through the recognition, definition, and investigation of Long COVID as a distinct clinical entity. Most notably, it highlights the transition from patient-reported symptoms to formal medical recognition and research initiatives. Each milestone is supported by peer-reviewed publications and official organizational declarations, demonstrating the scientific community's response to this emerging health challenge.

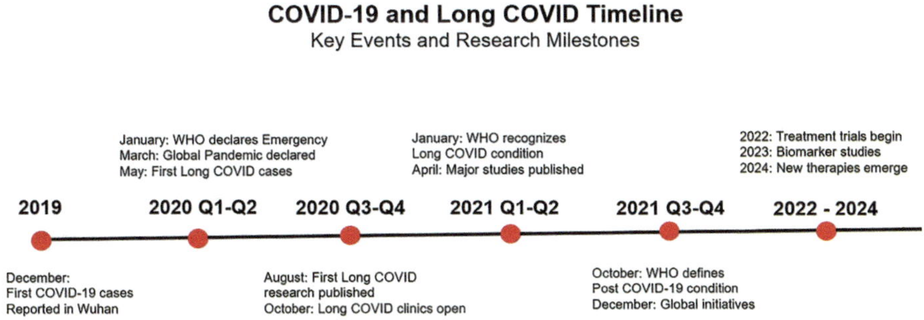

Fig. 1.3 Evolution of COVID-19 and Long COVID: Key events and research milestones (2019–2024). (Source: Compiled from WHO official statements, peer-reviewed literature, and major research initiatives. Key references include Wu et al. [5], Perego et al. [6], Carfì et al. [7], and Al-Aly et al. [8]). Abbreviations: WHO—World Health Organization

With this historical context established, we can better understand the current prevalence and impact of Long COVID, particularly its most common and debilitating manifestation: fatigue.

Understanding how Long COVID develops and manifests is crucial for recognizing and managing this condition. The progression from acute COVID-19 to Long COVID follows a distinct temporal pattern, with symptoms clustering into recognizable categories (Fig. 1.4). This transition diagram illustrates why symptoms may persist or emerge well after the initial infection has resolved, as documented by the PHOSP-COVID Collaborative Group [10].

The impact of Long COVID manifests through both its diverse symptomatology and significant economic consequences (Fig. 1.5). Among the constellation of symptoms, fatigue emerges as the predominant manifestation, affecting three-quarters of those with Long COVID.

The impact of Long COVID can be measured through its most prevalent symptoms and its substantial economic consequences. According to comprehensive studies published in The Lancet by the PHOSP-COVID Collaborative Group [10], fatigue emerges as the most pervasive symptom, affecting approximately three-quarters of Long COVID patients. This debilitating exhaustion fundamentally differs from everyday tiredness, often proving resistant to rest and significantly impacting daily activities [4, 11]. Following closely, cognitive difficulties—commonly referred to as "brain fog"—affect about 60% of patients, manifesting as difficulties with memory, concentration, and executive function as documented by the PHOSP-COVID Collaborative Group [9] in their prospective observational study of hospitalized UK patients.

Respiratory challenges persist in more than half of Long COVID cases, with 55% of patients reporting ongoing shortness of breath that can limit physical activity and exercise tolerance [12]. In their comprehensive review of post-acute COVID-19 syndrome, Nalbandian and colleagues [12] found that sleep disturbances affect 45% of individuals, creating a complex cycle where poor sleep quality often exacerbates other symptoms. Joint pain rounds out the top five most common symptoms, impacting approximately one-third of Long COVID patients and further contributing to reduced mobility and quality of life [10].

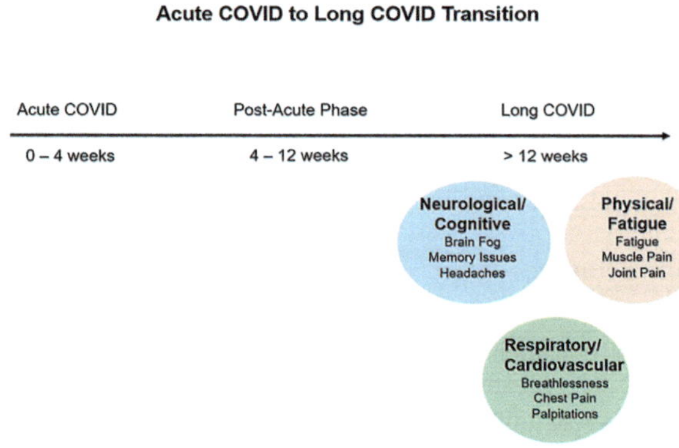

Fig. 1.4 Transition from acute COVID to Long COVID and major symptom clusters

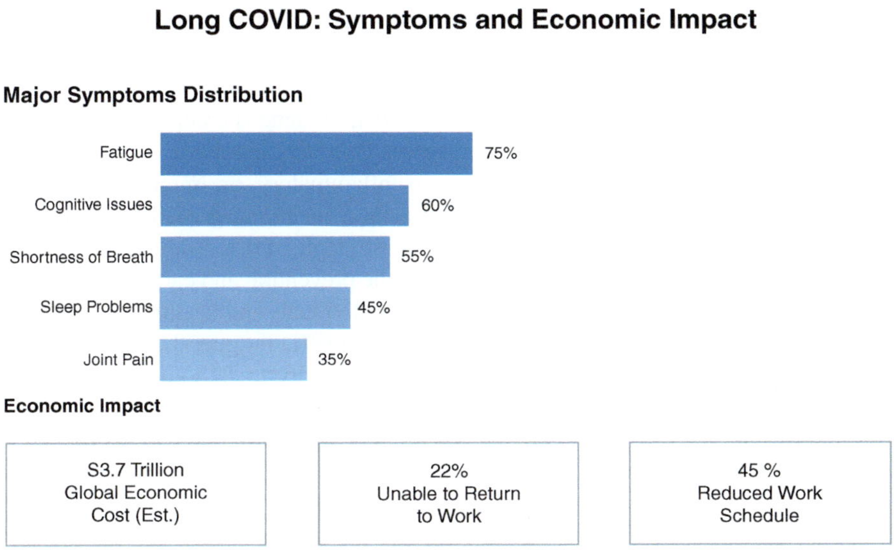

Fig. 1.5 Long COVID: Distribution of major symptoms and economic impact. (Sources: Hodges and Broadley [11], Nalbandian et al. [12])

The economic ramifications of Long COVID have proven equally significant, as highlighted in The Lancet's 2023 editorial "Long COVID research: unraveling a complex condition." Current estimates place the global economic burden at approximately $3.7 trillion, encompassing both direct healthcare costs and indirect impacts on productivity. The Lancet (2023) reports that more than one-fifth of Long COVID patients find themselves unable to return to work at all, while nearly half must modify their work schedules to accommodate their symptoms. This reduced workforce participation creates ripple effects throughout the economy, affecting not only individual households but entire industries and economic systems [10, 11].

The impact of Long COVID fatigue extends beyond the individual, with approximately one in ten affected individuals forced to stop working, leading to substantial economic losses [11]. A recent UK study [9] examined over 2000 patients hospitalized with COVID-19 and found that the proportion of patients reporting full recovery remained relatively consistent between the 5-month (26%) and one-year (29%) assessments. The study highlighted female sex and obesity as major risk factors for not achieving full recovery at the one-year mark, with these factors associated with more severe and ongoing health impairments, including reduced exercise performance and health-related quality of life.

The pathophysiology of Long COVID fatigue remains largely enigmatic, necessitating further research to elucidate its underlying mechanisms. Some hypotheses suggest that persistent inflammation, immune dysregulation, and mitochondrial dysfunction may play a role in the development of chronic fatigue following COVID-19 infection [13]. Additionally, the similarities between Long COVID fatigue and other post-viral fatigue syndromes, such as myalgic encephalomyelitis/

chronic fatigue syndrome (ME/CFS), suggest that insights gained from studying Long COVID may have broader implications for understanding and managing these conditions [13].

A comprehensive understanding of Long COVID fatigue is crucial for developing effective management and treatment strategies. This may involve a combination of pharmacological interventions, such as anti-inflammatory or immunomodulatory medications, and non-pharmacological approaches, including graded exercise therapy, cognitive-behavioral therapy, and pacing strategies [14]. Additionally, public health policies and messaging should prioritize the prevention and early identification of Long COVID fatigue to reduce its impact on individuals and healthcare systems.

This book focuses on the prevalence of Long COVID fatigue, the array of clinical symptoms it encompasses, and the urgent necessity to fathom its underlying pathogenic mechanisms. The objective is to enhance awareness about this debilitating manifestation of Long COVID as researchers work diligently to decipher the intricate puzzle of chronic post-viral fatigue, ultimately leading to improved management, treatment, and prevention strategies for those affected worldwide.

Throughout this book, the term "Long COVID fatigue" is used consistently. Some figures and references from published literature may use alternative terminology such as "post-COVID-19 fatigue," which should be considered equivalent for the purposes of this text.

Summary

This introductory chapter provides an examination of Long COVID and its development as a significant global health challenge, with particular focus on fatigue as its most prevalent symptom. Beginning with the WHO's declaration of COVID-19 as a pandemic, the chapter traces how our understanding of Long COVID has evolved over the past 4 years, affecting an estimated 65 million people worldwide.

The chapter begins by detailing the substantial societal and economic impact of Long COVID, revealing its far-reaching implications beyond individual health outcomes. With approximately 400 million cases globally by the end of 2023, representing 6–7% of adults and 1% of children affected, the condition has created profound ripple effects throughout society and the global economy. The chapter presents compelling evidence of the economic toll, estimating global costs at $1 trillion annually, equivalent to roughly 1% of global GDP. This impact is particularly evident in workforce participation, with studies showing affected individuals being 10% less likely to be employed and working significantly reduced hours when employed.

The chapter provides a thorough historical timeline, starting from the first COVID-19 cases in December 2019 through to current developments in 2024. It emphasizes the crucial role that patient advocacy played in the initial recognition of Long COVID, leading to its formal acknowledgment by the medical community. The chapter outlines key milestones in this progression, from the first research publications to the establishment of specialized clinics and the WHO's formal definition of the condition.

The chapter presents substantial epidemiological data, noting that while 10–20% of COVID-19 cases develop into Long COVID, the impact of fatigue is particularly significant, affecting approximately three-quarters of Long COVID patients. The text details how this fatigue differs from normal tiredness, being characterized by profound exhaustion that doesn't improve with rest and significantly impacts daily activities.

The economic implications are extensively discussed, with the global economic burden estimated at $3.7 trillion. The chapter highlights how Long COVID affects workforce participation, with about one-fifth of patients unable to return to work and nearly half requiring modified work schedules. This economic impact extends beyond individual households to affect entire industries and economic systems.

The chapter also explores the pathophysiological aspects of Long COVID fatigue, acknowledging that while the exact mechanisms remain unclear, current hypotheses point to persistent inflammation, immune dysregulation, and mitochondrial dysfunction as potential contributors. It draws parallels between Long COVID fatigue and other post-viral fatigue syndromes, suggesting that insights gained from studying Long COVID may have broader implications for understanding similar conditions.

The text concludes by emphasizing the need for both immediate management strategies and continued research into effective treatments, setting up the framework for more detailed discussions in subsequent chapters. Throughout, the chapter maintains a balanced perspective between scientific rigor and practical clinical applications, providing a comprehensive foundation for understanding this significant public health challenge.

References

1. Rudroff T, Fietsam AC, Deters JR, Bryant AD, Kamholz J. Post-COVID-19 fatigue: potential contributing factors. Brain Sci. 2020;10(12):1012.
2. Al-Aly Z, Davis H, McCorkell L, Soares L, Wulf-Hanson S, Iwasaki A, Topol EJ. Long COVID science, research and policy. Nat Med. 2024;30(8):2148–64. https://doi.org/10.1038/s41591-024-03173-6. Epub 2024 Aug 9
3. Sudre CH, Murray B, Varsavsky T, Graham MS, Penfold RS, Bowyer RC, et al. Attributes and predictors of long COVID. Nat Med. 2021;27(4):626–31.
4. Huang C, Huang L, Wang Y, Li X, Ren L, Gu X, et al. 6-month consequences of COVID-19 in patients discharged from hospital: a cohort study. Lancet. 2021;397(10270):220–32.
5. Wu F, Zhao S, Yu B, Chen YM, Wang W, Song ZG, et al. A new coronavirus associated with human respiratory disease in China. Nature. 2020;579(7798):265–9.
6. Perego E, Callard F, Stras L, Melville-Jóhannesson B, Pope R, Alwan NA. Why the patient-made term 'Long Covid' is needed. Welcome Open Res. 2020;5:224.

7. Carfì A, Bernabei R, Landi F. Persistent symptoms in patients after acute COVID-19. JAMA. 2020;324(6):603–5.
8. Al-Aly Z, Xie Y, Bowe B. High-dimensional characterization of post-acute sequelae of COVID-19. Nature. 2021;594(7862):259–64.
9. Su Y, Yuan D, Chen DG, Ng RH, Wang K, Choi J, et al. Multiple early factors anticipate post-acute COVID-19 sequelae. Cell. 2023;185(5):881–95.
10. PHOSP-COVID Collaborative Group. Clinical characteristics with inflammation profiling of long COVID and association with 1-year recovery following hospitalization in the UK: a prospective observational study. Lancet Respir Med. 2022;10(8):761–75.
11. Hodgson CL, Broadley T. Long COVID-unravelling a complex condition. Lancet Respir Med. 2023;11(8):667–8.
12. Nalbandian A, Sehgal K, Gupta A, Madhavan MV, McGroder C, Stevens JS, et al. Post-acute COVID-19 syndrome. Nat Med. 2021;27(4):601–15.
13. Komaroff AL, Lipkin WI. Insights from myalgic encephalomyelitis/chronic fatigue syndrome may help unravel the pathogenesis of post-acute COVID-19 syndrome. Trends Mol Med. 2021;27(9):895–906.
14. Décary S, Gaboury I, Poirier S, Garcia C, Simpson S, Bull M, et al. Humility and acceptance: working within our limits with long COVID and Myalgic encephalomyelitis/chronic fatigue syndrome. J Orthop Sports Phys Ther. 2021;51(5):197–200.

Understanding Fatigue 2

2.1 What Is Fatigue?

Fatigue remains poorly comprehended despite substantial efforts to clarify its pathogenic mechanisms. Potentially, this stems from an inability to attribute fatigue to a sole origin. Alterations in neurotransmitters, inflammation, psychological conditions, stress, cognitive dysfunction, and substrate availability are possible factors contributing to fatigue. Here, we expand on a prior model of fatigue to delineate elements specifically underlying Long COVID fatigue. This framework is intended to provide researchers and clinicians greater insight into fatigue determinants and suggest further research directions.

Numerous fatigue investigations fail to definitively characterize it with definitions varying extensively. Moreover, fatigue roots diverge across diseases. Most disease research overlooks discriminating fatigue from confounding occurrences like mood disorders or sleep disturbances while discounting comorbidities as contributors. Consequently, Rudroff et al. [1] recently proposed a unified taxonomy for neurological fatigue applicable to post-COVID-19 fatigue. *Post-COVID-19 fatigue is defined as decreased physical and/or mental performance stemming from COVID-19-induced changes in central, psychological, and/or peripheral factors.*

In the following chapters, it is discussed how these physiological factors may interact with task, environment, and performance capacity to further impact Long COVID fatigue (Fig. 2.1).

T. Rudroff, *Long COVID Fatigue*, https://doi.org/10.1007/978-3-031-89474-9_2

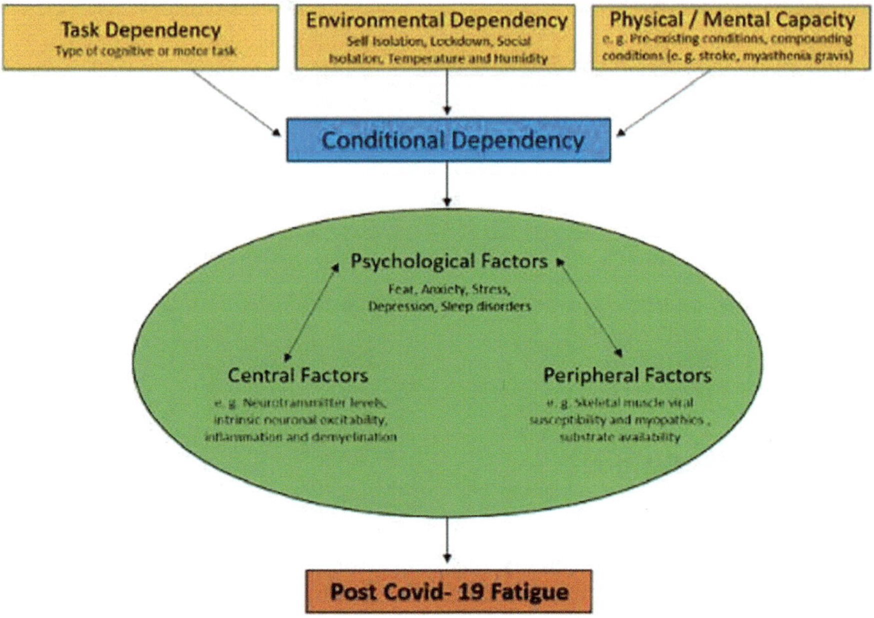

Fig. 2.1 Conceptual framework of fatigue: Interactive relationships between physiological factors and conditional dependencies. Note: The term "post-COVID-19 fatigue" used in this published figure is equivalent to "Long COVID fatigue" used throughout this book

2.2 Fatigue and Fatigability: Understanding the Distinction

The distinction between fatigue and fatigability represents a crucial consideration in post-COVID-19 research, as highlighted by recent studies examining these phenomena. According to Enoka et al. [2], fatigue manifests as a complex interplay of various physiological and psychological factors, derived from interoceptive feedback that reflects the body's homeostatic state. This subjective experience encompasses feelings of low energy, tiredness, decreased motivation, and difficulties in concentration, which can only be evaluated through self-reported measures.

The concept of fatigability, however, is further divided into two distinct components: perceived fatigability and objective fatigability. As demonstrated in recent research by Fietsam et al. [3], perceived fatigability involves the subjective estimation of an individual's past or future work capacity. In contrast, objective fatigability is measured through quantifiable changes in performance metrics following specific tasks. This differentiation has proven particularly significant in post-COVID-19 research, where studies have revealed that perceived fatigability and fatigue are more prevalent than objective fatigability among affected individuals.

The importance of this distinction becomes especially apparent when examining clinical findings. For instance, Fietsam et al. [3] conducted a comparative study of fatigue, perceived fatigue, and objective fatigability in post-COVID-19 patients with and without reported persistent fatigue symptoms. Their results revealed an intriguing pattern: while subjects with fatigue symptoms reported worse levels of perceived and fatigability fatigue, they showed no significant difference in performance on an isokinetic fatigue task (objective fatigability measure) compared to those without fatigue symptoms.

This pattern mirrors observations in other conditions, particularly multiple sclerosis. DeLuca et al. [4] proposed that despite showing comparable performance levels on fatigue tasks (objective fatigability) to healthy subjects, increased perceived fatigability might result from higher cerebral energy demands required to accomplish the same task. This hypothesis has led to new research directions, with Rudroff et al. [5] suggesting the use of positron emission tomography (PET) imaging as a crucial tool for identifying underlying mechanisms driving brain changes in individuals with post-COVID-19 fatigue symptoms.

Understanding this distinction has significant implications for both research methodology and clinical practice. Rudroff [6] emphasizes that future investigations should employ diverse assessment approaches, including neuroimaging procedures such as functional magnetic resonance imaging (fMRI) and fluorodeoxyglucose-positron emission tomography (FDG-PET) to measure changes in cerebral blood flow and glucose metabolism. Additionally, psychological factors should be evaluated through neuropsychological tests, while peripheral factors can be assessed by comparing variations in electromyography signals with maximal force or power output.

The differentiation between fatigue and fatigability is not merely academic; it has practical implications for patient care and treatment strategies. As Kluger et al. [7] noted in their proposal for a unified taxonomy of fatigue in neurologic illnesses, this distinction helps clinicians and researchers better understand the multifaceted nature of fatigue symptoms and develop more targeted interventions. This understanding becomes particularly crucial in the context of post-COVID-19 syndrome, where the complexity of symptoms and their varying manifestations require precise diagnostic and therapeutic approaches.

This nuanced understanding of fatigue and fatigability continues to shape research directions in post-COVID-19 studies, emphasizing the need for comprehensive assessment methods that account for both subjective experiences and objective measurements. As our knowledge of post-COVID-19 syndrome evolves, this distinction will likely play an increasingly important role in developing effective treatment strategies and understanding the long-term implications of the condition.

2.3 Comparison of Post-viral Fatigue Syndromes

To better understand the unique characteristics of Long COVID fatigue, it is instructive to compare it with other well-documented post-viral syndromes (Fig. 2.2). This comparison reveals several distinctive features of Long COVID that help explain its particular challenges and treatment requirements.

Comparison of Post-Viral Fatigue Syndromes

	Long COVID	Post-SARS (2003)	Post-EBV	ME/CFS
Onset Pattern	• Variable: mild to severe cases • Can occur immediately or develop weeks after infection	• Usually after severe cases • Clear link to acute illness • Gradual onset post-recovery	• Following acute EBV • Often after mononucleosis • Gradual development	• Often post-viral trigger • Can be gradual or sudden • Multiple possible triggers
Key Symptoms	• Multi-system involvement • Prominent brain fog • Cardiovascular symptoms	• Respiratory focus • Muscular weakness • Less cognitive impact	• Profound fatigue • Lymphatic symptoms • Sore throat common	• Post-exertional malaise • Cognitive difficulties • Sleep disturbances
Duration	• Highly variable • Months to years • Unpredictable course	• 6 months to 2 years • More predictable course • Better defined phases	• 3-6 months typical • Can extend longer • Generally self-limiting	• Often chronic (years) • Relapsing-remitting • Long-term condition
Mechanisms	• Persistent inflammation • Autoimmune components • Vascular dysfunction	• Post-viral inflammation • Respiratory damage • Muscle deconditioning	• Viral reactivation • Immune dysregulation • Lymphatic involvement	• Energy metabolism issues • Autonomic dysfunction • Immune abnormalities
Recovery Pattern	• Variable recovery • Frequent relapses • Multiple treatments needed	• Gradual improvement • Better prognosis • Defined recovery phases	• Generally good recovery • Predictable course • Can have reactivation	• Often chronic course • Variable prognosis • Complex management

Fig. 2.2 A systematic comparison of Long COVID fatigue with three well-documented post-viral syndromes: post-SARS (2003), post-EBV (Epstein-Barr Virus), and ME/CFS (myalgic encephalomyelitis/chronic fatigue syndrome)

The distinctive characteristics of Long COVID compared to other post-viral syndromes reveal important insights into its unique pathophysiology and clinical challenges. Drawing from multiple landmark studies, several key differences emerge in onset patterns, symptom manifestation, disease trajectory, and recovery characteristics.

The onset pattern of Long COVID demonstrates remarkable variability, as documented by the PHOSP-COVID Collaborative Group [8] in their comprehensive study published in The Lancet Respiratory Medicine. Unlike post-SARS syndrome, which predominantly follows severe acute illness [9], Long COVID can develop after both mild and severe cases. This contrasts sharply with post-EBV syndrome, which typically follows a clear bout of infectious mononucleosis with predictable progression [10], and ME/CFS, which often has a more defined triggering event.

The symptom profile of Long COVID presents unique challenges due to its broad multi-system involvement. Nalbandian et al. [11] in Nature Medicine highlighted distinctive features, including prominent cardiovascular manifestations and cognitive dysfunction ("brain fog"), setting it apart from the predominantly respiratory focus of post-SARS syndrome [12]. While post-EBV syndrome typically centers on fatigue and lymphatic symptoms [13], and ME/CFS is characterized by post-exertional malaise [14], Long COVID presents a more diverse and variable symptom pattern.

Duration and disease trajectory show marked differences as well. The PHOSP-COVID Collaborative Group [8] documented highly variable recovery timeframes in Long COVID patients, contrasting with the more predictable 6-month to 2-year course observed in post-SARS cases [15]. Post-EBV syndrome typically resolves within 3–6 months [16], while ME/CFS tends toward chronicity but with more predictable patterns [17].

The underlying mechanisms of Long COVID appear uniquely complex. Kuchler et al. [15] identified distinctive patterns of persistent inflammation and vascular dysfunction, differing from the focused respiratory damage in post-SARS [12] and the viral reactivation patterns typical of post-EBV syndrome [18]. ME/CFS, while sharing some features, primarily involves energy metabolism and autonomic dysfunction [17].

Recovery patterns in Long COVID show remarkable heterogeneity. The PHOSP-COVID Collaborative Group [8] documented variable recovery trajectories and frequent relapses, requiring multiple concurrent treatment approaches. This differs significantly from the more gradual but predictable improvement seen in post-SARS [9], the generally favorable outcomes in post-EBV syndrome [13], and the more stable, though chronic, course of ME/CFS [17].

These fundamental differences highlight why standardized post-viral treatment protocols may prove insufficient for Long COVID patients. As emphasized by The Lancet [13] in their editorial on Long COVID research, the condition's unique characteristics necessitate individualized, comprehensive treatment approaches that address its multi-system nature and variable presentation.

This comparison not only enhances our understanding of Long COVID's distinct pathophysiology but also underscores the importance of developing targeted therapeutic strategies that acknowledge its unique characteristics. Future research, as suggested by the PHOSP-COVID Collaborative Group [8], should focus on understanding these distinguishing features to develop more effective, personalized treatment approaches.

Building upon the conceptual framework introduced in Chap. 2 (Fig. 2.1), Chap. 3 provides a detailed examination of the factors contributing to Long COVID fatigue. As illustrated in Fig. 2.1, fatigue manifests through the complex interaction of central, psychological, and peripheral factors, all of which operate within a framework of conditional dependencies including task demands, environmental conditions, and individual capabilities. Understanding these components and their interactions is crucial for developing effective treatment strategies for Long COVID fatigue. Each factor represents a potential therapeutic target, and their interdependent nature suggests that comprehensive treatment approaches may be necessary. The following sections explore each of these factors in detail, examining the current evidence for their role in Long COVID fatigue and highlighting areas requiring further research.

Summary

This comprehensive chapter provides a detailed exploration of post-COVID-19 fatigue through three interconnected sections that build upon each other to create a thorough understanding of this complex condition. Beginning with the fundamental question "What is fatigue?", the chapter acknowledges the challenges in comprehending fatigue despite substantial research efforts, primarily due to its multifaceted origins involving neurotransmitter alterations, inflammation, psychological conditions, stress, cognitive dysfunction, and substrate availability.

The chapter then delves into the crucial distinction between fatigue and fatigability, a differentiation that proves particularly significant in post-COVID-19 research. Drawing from recent studies, it explains how fatigue manifests as a complex interplay of physiological and psychological factors, while fatigability encompasses both perceived and objective components. This distinction becomes especially relevant when examining clinical findings, where research has shown that perceived fatigability is more prevalent than objective fatigability in post-COVID-19 patients. The chapter emphasizes how this understanding has important implications for both research methodology and clinical practice, suggesting the need for diverse assessment approaches including neuroimaging, neuropsychological testing, and electromyography.

In its comparative analysis section, the chapter offers an instructive comparison between Long COVID fatigue and other post-viral syndromes, including post-SARS, post-EBV, and ME/CFS. This comparison reveals distinctive characteristics of Long COVID, particularly in its onset patterns, symptom manifestation, disease trajectory, and recovery characteristics. Unlike other post-viral conditions, Long COVID shows remarkable variability in its onset and can develop after both mild and severe cases. Its symptom profile presents unique challenges due to broad multi-system involvement, including distinctive cardiovascular manifestations and cognitive dysfunction.

The chapter concludes by introducing a conceptual framework that illustrates how fatigue manifests through the complex interaction of central, psychological, and peripheral factors, all operating within conditional dependencies including task demands, environmental conditions, and individual capabilities. This framework serves as a bridge to the subsequent detailed examination of these factors, emphasizing their role as potential therapeutic targets and the need for comprehensive treatment approaches.

Throughout the chapter, the authors maintain a strong focus on the practical implications of this understanding for patient care and treatment strategies. They emphasize that the unique characteristics of Long COVID necessitate individualized, comprehensive treatment approaches that address its multi-system nature and variable presentation. This thorough exploration not only enhances our understanding of Long COVID's distinct pathophysiology but also provides a foundation for future research directions aimed at developing more effective, personalized treatment approaches.

References

1. Rudroff T, Fietsam AC, Deters JR, Bryant AD, Kamholz J. Post-COVID-19 fatigue: potential contributing factors. Brain Sci. 2020;10(12):1012.
2. Enoka RM, Almuklass AM, Alenazy M, Alvarez E, Duchateau J. Distinguishing between fatigue and fatigability in multiple sclerosis. Neurorehabil Neural Repair. 2021;35(11):960–73. https://doi.org/10.1177/15459683211046257.
3. Fietsam AC, Bryant AD, Rudroff T. Fatigue and perceived fatigability, not objective fatigability, are prevalent in people with post-COVID-19. Exp Brain Res. 2022;3:1–9. https://doi.org/10.1007/s00221-022-06518-0.
4. DeLuca J, Genova HM, Hillary FG, Wylie G. Neural correlates of cognitive fatigue in multiple sclerosis using functional MRI. J Neurol Sci. 2008;270(1–2):28–39. https://doi.org/10.1016/j.jns.2008.01.018.
5. Rudroff T, Workman CD, Boles Ponto LL. 18F-FDG-PET imaging for post-COVID-19 brain and skeletal muscle alterations. Viruses. 2021;13(11):2283. https://doi.org/10.3390/v13112283.
6. Rudroff T. The challenges of post-COVID-19 fatigue research. Front Psychol. 2023;14:1120928. https://doi.org/10.3389/fpsyg.2023.1120928. PMID: 37051605; PMCID: PMC10083308.
7. Kluger BM, Krupp LB, Enoka RM. Fatigue and fatigability in neurologic illnesses: proposal for a unified taxonomy. Neurology. 2013;80(4):409–16. https://doi.org/10.1212/WNL.0b013e31827f07be.
8. PHOSP-COVID Collaborative Group. Clinical characteristics with inflammation profiling of long COVID and association with 1-year recovery following hospitalization in the UK: a prospective observational study. Lancet Respir Med. 2022;10(8):761–75.
9. Lam MH, Wing YK, Yu MW, Leung CM, Ma RC, Kong AP, et al. Mental morbidities and chronic fatigue in severe acute respiratory syndrome survivors: long-term follow-up. Arch Intern Med. 2009;169(22):2142–7.
10. Hickie I, Davenport T, Wakefield D, Vollmer-Conna U, Cameron B, Vernon SD, et al. Post-infective and chronic fatigue syndromes precipitated by viral and non-viral pathogens: prospective cohort study. BMJ. 2006;333(7568):575.
11. Nalbandian A, Sehgal K, Gupta A, Madhavan MV, McGroder C, Stevens JS, et al. Post-acute COVID-19 syndrome. Nat Med. 2021;27(4):601–15.
12. Hui DS, Joynt GM, Wong KT, Gomersall CD, Li TS, Antonio G, et al. Impact of severe acute respiratory syndrome (SARS) on pulmonary function, functional capacity and quality of life in a cohort of survivors. Thorax. 2005;60(5):401–9.
13. White PD, Goldsmith KA, Johnson AL, Potts L, Walwyn R, DeCesare JC, et al. Comparison of adaptive pacing therapy, cognitive behavior therapy, graded exercise therapy, and specialist medical care for chronic fatigue syndrome (PACE): a randomized trial. Lancet. 2011;377(9768):823–36.
14. Ahmed H, Patel K, Greenwood DC, Halpin S, Lewthwaite P, Salawu A, et al. Long-term clinical outcomes in survivors of severe acute respiratory syndrome and Middle East respiratory syndrome coronavirus outbreaks after hospitalization or ICU admission: a systematic review and meta-analysis. J Rehabil Med. 2020;52(5):1–11.
15. Kuchler T, Günthner R, Ribeiro A, Hausinger R, Streese L, Wöhnl A, et al. Persistent endothelial dysfunction in post-COVID-19 syndrome and its associations with symptom severity and chronic inflammation. Angiogenesis. 2023;26(4):547–63.
16. Katz BZ, Shiraishi Y, Mears CJ, Binns HJ, Taylor R. Chronic fatigue syndrome after infectious mononucleosis in adolescents. Pediatrics. 2009;124(1):189–93.
17. Institute of Medicine. Beyond myalgic encephalomyelitis/chronic fatigue syndrome: redefining an illness. Washington, DC: National Academies Press; 2015.
18. Cameron B, Bharadwaj M, Burrows J, Fazou C, Wakefield D, Hickie I, et al. Prolonged illness after infectious mononucleosis is associated with altered immunity but not with increased viral load. J Infect Dis. 2006;193(5):664–71.

3 Factors Contributing to Long COVID Fatigue

3.1 Central Factors

It remains ambiguous whether COVID-19 is neuroinvasive. However, emerging evidence suggests that SARS-CoV-2, like other human coronaviruses, may utilize multiple pathways to permeate the central nervous system (CNS) [1–3]. Recent research has identified three primary routes of CNS entry: the neural pathway through the olfactory system, the hematogenous pathway via blood circulation and ACE2 receptor binding, and the lymphatic pathway through the brain's glymphatic system [4–6].

Figure 3.1 illustrates these pathways and their downstream effects leading to fatigue:

The neural pathway involves initial infection of the olfactory epithelium followed by retrograde axonal transport, explaining the frequent occurrence of anosmia in COVID-19 patients [7]. Through the hematogenous route, the virus binds to ACE2 receptors on blood-brain barrier endothelial cells, compromising barrier integrity and allowing viral particles and inflammatory mediators to enter the CNS [8]. The lymphatic pathway, utilizing the brain's waste clearance system, provides an additional route for viral entry through meningeal lymphatic vessels [9].

Once in the CNS, the virus triggers three major pathophysiological processes that contribute to fatigue development. First, it initiates neuroinflammation through microglial activation and cytokine release [10]. Second, it disrupts key neurotransmitter systems, including dopamine and serotonin, affecting motivation, mood, and energy levels [11–14]. Third, it induces structural changes, including demyelination and altered brain connectivity patterns [15].

Neuroimaging studies are beginning to demonstrate these central fatigue factors in COVID-19. Using ^{18}F-fluorodeoxyglucose-positron emission tomography (FDG-PET), Delorme et al. [11] found frontal hypometabolism and cerebellar hypermetabolism in fatigued COVID-19 patients, potentially impacting fatigue. While frontal hypometabolism consistently associates with fatigue in neurological disorders [16–18], depression may also cause these changes. Another FDG-PET study

T. Rudroff, *Long COVID Fatigue*, https://doi.org/10.1007/978-3-031-89474-9_3

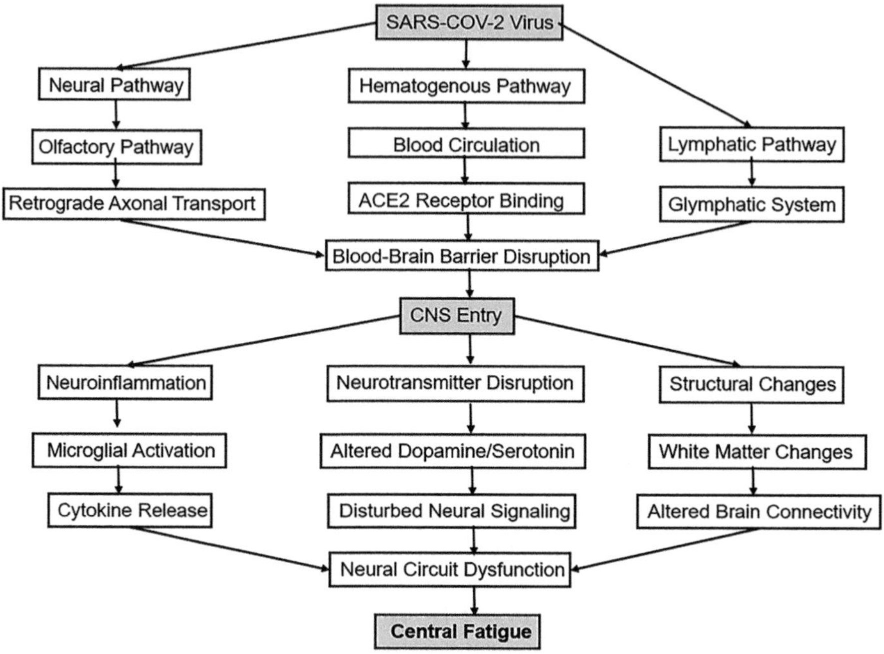

Fig. 3.1 SARS-CoV-2 brain entry pathways and fatigue development

[12] revealed widespread cerebral hypometabolism in COVID-19 patients, needing longer follow-up to clarify links to persistent fatigue.

The convergence of these pathways—viral invasion, neuroinflammation, neurotransmitter disruption, and structural changes—creates a complex cascade leading to neural circuit dysfunction and central fatigue. Moreover, pandemic-related lockdowns have normalized extensive physical inactivity periods, potentially decreasing motor neuron excitability [19, 20]. Together with COVID-19-induced motor unit conduction velocity reductions seen on nerve studies, this may further enhance fatigue [21, 22].

Understanding these interconnected mechanisms is crucial for developing targeted therapeutic approaches for post-COVID fatigue syndrome. The persistence of these changes may explain the chronic nature of fatigue in Long COVID and its association with ongoing neurological symptoms.

3.2 Psychological Factors

The psychological dimension of COVID-19-related fatigue presents a complex interplay between neurobiological changes and psychological responses to the pandemic experience. For numerous patients, COVID-19-linked fatigue manifests alongside a constellation of psychological symptoms, including pervasive stress, anxiety, depression, and fear [21]. These psychological manifestations appear to have both direct viral-mediated mechanisms and indirect psychosocial causes, creating a multifaceted challenge for both patients and healthcare providers.

Figure 3.2 illustrates the complex interactions between direct viral effects, indirect psychosocial impacts, and their contributions to fatigue development. The flowchart demonstrates how COVID-19 affects psychological well-being through both neurobiological mechanisms and psychosocial pathways, creating a self-perpetuating cycle of fatigue and psychological distress.

The implementation of public health measures, while crucial for controlling viral transmission, has introduced additional psychological burdens. Quarantines, social distancing requirements, and isolation periods effectively slow viral spread but may adversely intensify fatigue in recovering patients [22–24]. Research has documented a range of psychological impacts, including post-traumatic stress symptoms, cognitive confusion, emotional dysregulation, and heightened anger responses. These psychological sequelae appear to contribute significantly to the overall fatigue burden [10].

The indirect effects cascade through multiple psychosocial pathways. Social isolation, mandated by public health measures, triggers a complex stress response involving the hypothalamic-pituitary-adrenal (HPA) axis. As documented in systematic reviews, prolonged isolation can lead to elevated stress responses and inflammatory markers, creating a physiological environment conducive to fatigue development [15]. This stress response interacts synergistically with anxiety and fear about the illness itself.

Particularly significant is the perpetuating cycle illustrated in the flowchart, where fatigue itself becomes a source of psychological distress. The inflammatory

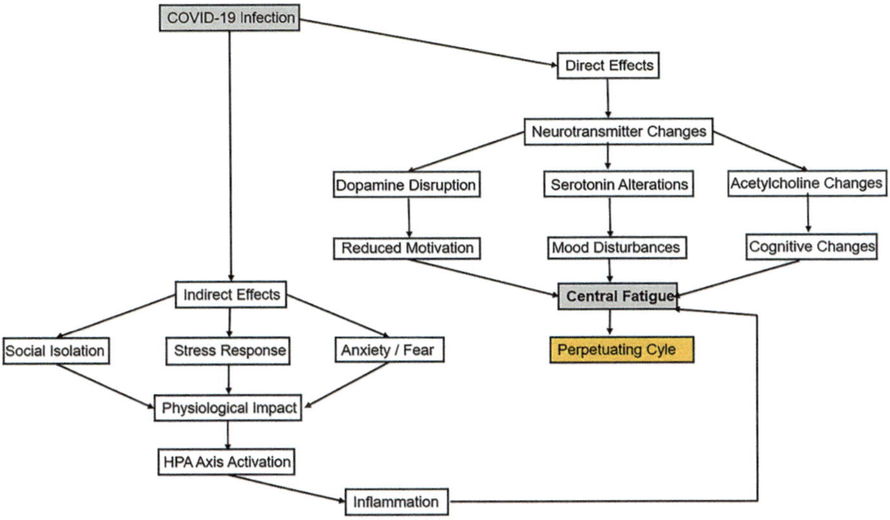

Fig. 3.2 Psychological factors contributing to COVID-19 related fatigue

response, triggered by both direct viral effects and psychological stress, serves as a critical mediator in this cycle [25]. Studies have shown that this inflammatory response can persist even in mildly affected patients, potentially contributing to ongoing psychological symptoms and fatigue.

Clinical research has demonstrated how this complex interplay manifests in treatment outcomes, suggesting that effective interventions must address both the neurobiological and psychological aspects of fatigue [26]. The bidirectional arrows in the flowchart emphasize how each component can influence and be influenced by others, creating a dynamic system that requires comprehensive therapeutic approaches.

This integrated understanding has led to the development of novel therapeutic approaches that combine pharmacological interventions with psychological support, addressing the indirect effects of the illness. Understanding the bidirectional relationship between psychological factors and fatigue is crucial for developing effective interventions [14]. These therapeutic approaches should address both the neurobiological substrates of fatigue and its psychological manifestations through combined treatments and psychological interventions such as cognitive behavioral therapy and stress management techniques [27].

3.3 Peripheral Factors

Long COVID fatigue manifests through complex peripheral mechanisms that extend beyond central nervous system involvement. These peripheral pathways encompass multiple organ systems and metabolic processes, creating a multifaceted contribution to fatigue development. Understanding these mechanisms is crucial for developing targeted therapeutic approaches and improving patient outcomes.

Figure 3.3 illustrates the intricate network of peripheral mechanisms contributing to Long COVID fatigue:

SARS-CoV-2 infection can affect multiple peripheral tissues, with significant impact on musculoskeletal function. Disser and colleagues have documented extensive musculoskeletal consequences of COVID-19, including muscle damage,

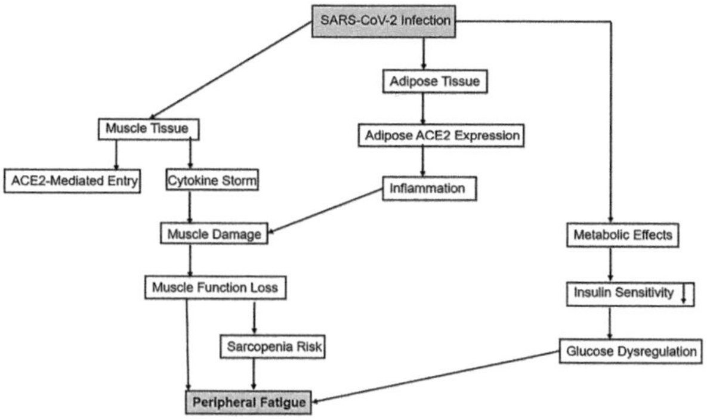

Fig. 3.3 Peripheral mechanisms contributing to Long COVID fatigue

weakness, and reduced exercise tolerance [28]. The virus's ability to affect multiple organ systems creates a complex pattern of peripheral manifestations that contribute to fatigue development.

The role of adipose tissue and metabolic dysfunction has emerged as a significant factor in COVID-19 outcomes. Systematic review evidence indicates that obesity is associated with poor COVID-19 outcomes, suggesting that adipose tissue may serve as both a viral reservoir and a source of inflammatory mediators [29]. This metabolic component may help explain the prolonged nature of fatigue in some patients.

A prolonged inflammatory response has been documented even in mildly affected patients, as demonstrated by Doykov et al. [25]. This persistent inflammation may contribute to ongoing muscle dysfunction and fatigue. The systemic nature of this inflammatory response appears to create a cycle of tissue dysfunction and fatigue that can persist well beyond the acute phase of infection.

The peripheral manifestations of COVID-19 often interact with neurological symptoms, creating a complex clinical picture. The Global COVID-19 Neuro Research Coalition has documented a wide range of neurological manifestations among COVID-19 patients [30], suggesting that peripheral and central mechanisms often work in concert to produce fatigue symptoms.

Mehandru and Merad [31] have described the pathological sequelae of long-haul COVID-19, highlighting how persistent tissue dysfunction can lead to chronic fatigue. This understanding is crucial for developing targeted interventions. The involvement of the nervous system in SARS-CoV-2 infection, as described by Li et al. [32], provides additional insight into how peripheral and central mechanisms might interact to produce fatigue.

The concept of post-acute COVID-19 syndrome, thoroughly documented by Nalbandian et al. [26], encompasses many of these peripheral mechanisms. Their work suggests that understanding these peripheral pathways is essential for developing effective treatments for Long COVID fatigue. The persistence of symptoms appears to involve ongoing inflammation and tissue dysfunction, creating a challenging clinical scenario that requires comprehensive management approaches.

This complex interplay of peripheral mechanisms highlights the need for multifaceted treatment approaches that address both the immediate and long-term consequences of COVID-19 infection. Future research should focus on understanding the relative contributions of these different mechanisms to develop more targeted therapeutic interventions.

3.4 Conditional Dependency

Conditional reliance denotes that central, peripheral, and psychological fatigue factors depend on task, environment, and individual abilities (Fig. 2.1). Assignments influence fatigue in healthy and diseased subjects [33–35], affirming task-dependency. Rather than any sole determinant, human studies show fatigue stems from interacting cognitive and motor demands. Environmental factors like temperature and humidity also greatly impact neurological fatigue [36]. Beyond health and financial burdens, lockdowns and isolation may negatively affect physical and mental capacity [27], while pandemic-related inactivity and distress can further enhance fatigue. Pre-existing conditions factor into capacity as well [37]. Comorbidities like cardiovascular illness, hypertension, and COPD feature shared fatigue symptoms. Their occurrence may disproportionately perpetuate Long COVID fatigue versus healthy counterparts.

Here, fatigue definitions and contributing factors in Long COVID fatigue are proposed. This remains incomplete, with this hypothetical model necessitating further research to explain all mechanisms and validate proposals. Studies should concentrate on discrete fatigue outcome measures. Given multifaceted COVID-19 fatigue, diverse methods targeting distinct influential factors are imperative in future studies. Currently, numerous techniques can evaluate many factors, which will be described in the following chapters. Neuroimaging like FDG-PET can assess central glucose metabolic shifts [38–41]. Neuropsychological testing can gauge psychological facets. Peripheral factors come from electromyography signal comparisons to maximal force/power [42].

3.5 Integrative Understanding of Long COVID Fatigue

The intricate interplay between central, peripheral, and psychological factors in Long COVID fatigue demonstrates that no single mechanism acts in isolation. Rather, these factors form an interconnected network where changes in one domain inevitably influence the others. As illustrated throughout this chapter, SARS-CoV-2 infection can simultaneously affect brain function through direct viral invasion, alter peripheral tissue metabolism, and trigger significant psychological responses. This complex pattern of dysfunction creates a self-perpetuating cycle where neurological, muscular, and emotional factors amplify and maintain fatigue symptoms.

The conditional dependency of these mechanisms on task demands, environmental conditions, and individual characteristics further emphasizes the multifaceted nature of Long COVID fatigue. A patient's experience of fatigue may vary significantly based on physical activity levels, emotional state, and pre-existing conditions. Understanding this complexity is crucial for both research and clinical practice, as it suggests that effective interventions must address multiple contributing factors simultaneously.

The challenge for future research lies in developing integrated assessment approaches that can capture the full spectrum of fatigue manifestations. While individual techniques like FDG-PET imaging, neuropsychological testing, and electromyography can provide valuable insights into specific aspects of fatigue, combining these methods may offer a more comprehensive understanding of how different factors interact and evolve over time.

This holistic perspective on Long COVID fatigue emphasizes the need for personalized, multimodal treatment strategies that consider the unique combination of central, peripheral, and psychological factors affecting each patient. Only by acknowledging and addressing this complexity can we hope to develop more effective interventions for this challenging condition.

Summary

This chapter examines the complex nature of Long COVID fatigue through three primary mechanisms: central, psychological, and peripheral factors, each contributing to and influencing the others in distinct ways.

The central mechanisms begin with SARS-CoV-2's potential neuroinvasion through three main pathways: neural (via the olfactory system), hematogenous (through blood circulation), and lymphatic (via the brain's glymphatic system). Once in the central nervous system, the virus triggers neuroinflammation, disrupts neurotransmitter systems, and causes structural changes in brain connectivity. Neuroimaging studies, particularly using FDG-PET, have revealed significant alterations in brain metabolism, especially in frontal regions associated with fatigue.

Psychological factors manifest through both direct viral-mediated mechanisms and indirect psychosocial impacts. The chapter illustrates how public health measures, while necessary, have contributed to psychological burden through isolation and stress responses. These psychological effects create a self-reinforcing cycle where fatigue and psychological distress amplify each other, mediated by inflammatory responses that can persist even after the acute infection resolves.

Peripheral mechanisms extend beyond the central nervous system, affecting multiple organ systems. The virus impacts musculoskeletal function, causes metabolic dysfunction, and triggers systemic inflammation. These peripheral effects often interact with central and psychological mechanisms, creating a complex web of symptoms that can persist long after the initial infection.

The chapter introduces the concept of conditional dependency, emphasizing how fatigue manifestations depend on task demands, environmental conditions, and individual characteristics. This understanding highlights the need for comprehensive assessment approaches that combine various diagnostic techniques, including neuroimaging, neuropsychological testing, and electromyography.

Future research directions should focus on developing integrated assessment approaches and personalized treatment strategies that address the multiple facets of Long COVID fatigue simultaneously. The chapter concludes by emphasizing that effective interventions must consider the unique combination of factors affecting each patient, as no single mechanism acts in isolation.

This holistic understanding of Long COVID fatigue provides a foundation for developing more effective, targeted interventions that address the condition's complex and interconnected nature.

References

1. Desforges M, Le Coupanec A, Dubeau P, Bourgouin A, Lajoie L, Dubé M, et al. Human coronaviruses and other respiratory viruses: underestimated opportunistic pathogens of the central nervous system? Viruses. 2020;12(1):14.
2. Talbot HK, Falsey AR. The diagnosis of viral respiratory disease in older adults. Clin Infect Dis. 2010;50(5):747–51.
3. Pezzini A, Padovani A. Lifting the mask on neurological manifestations of COVID-19. Nat Rev Neurol. 2020;16:636–44.
4. Meinhardt J, Radke J, Dittmayer C, Franz J, Thomas C, Mothes R, et al. Olfactory transmucosal SARS-CoV-2 invasion as a port of central nervous system entry in individuals with COVID-19. Nat Neurosci. 2021;24(2):168–75.
5. Rhea EM, Logsdon AF, Hansen KM, Williams LM, Reed MJ, Baumann KK, et al. The S1 protein of SARS-CoV-2 crosses the blood-brain barrier in mice. Nat Neurosci. 2021;24(3):368–78.
6. Song E, Zhang C, Israelow B, Lu-Culligan A, Prado AV, Skriabine S, et al. Neuroinvasion of SARS-CoV-2 in human and mouse brain. J Exp Med. 2021;218(3):e20202135.
7. Yang AC, Kern F, Losada PM, Agam MR, Maat CA, Schmartz GP, et al. Dysregulation of brain and choroid plexus cell types in severe COVID-19. Nature. 2021;595(7868):565–71.
8. Wenzel J, Lampe J, Müller-Fielitz H, Schuster R, Zille M, Müller K, et al. The SARS-CoV-2 main protease Mpro causes microvascular brain pathology by cleaving NEMO in brain endothelial cells. Nat Neurosci. 2021;24(11):1522–33.
9. Douaud G, Lee S, Alfaro-Almagro F, Arthofer C, Wang C, McCarthy P, et al. SARS-CoV-2 is associated with changes in brain structure in UK Biobank. Nature. 2022;604(7907):697–707.
10. Davis HE, Assaf GS, McCorkell L, Wei H, Low RJ, Re'em Y, et al. Characterizing long COVID in an international cohort: 7 months of symptoms and their impact. EClinicalMedicine. 2021;38:101019.
11. Delorme C, Paccoud O, Kas A, Hesters A, Bombois S, Shambrook P, et al. COVID-19-related encephalopathy: a case series with brain FDG-positron-emission tomography/computed tomography findings. Eur J Neurol. 2020;27(12):2651–7.
12. Guedj E, Campion JY, Dudouet P, Kaphan E, Bregeon F, Tissot-Dupont H, et al. 18F-FDG brain PET hypometabolism in patients with long COVID. Eur J Nucl Med Mol Imaging. 2021;48(9):2823–33.
13. Wood ER, Eid T, Carvajal AG, Anger HC, Reyland E, Hashemiyoon R, et al. Neurological manifestations of COVID-19: a systematic review and meta-analysis of proportions. Neurol Sci. 2021;42(10):3931–45.
14. Komaroff AL, Bateman L. Will COVID-19 lead to myalgic encephalomyelitis/chronic fatigue syndrome? Front Med (Lausanne). 2021;7:606824.
15. Proal AD, VanElzakker MB. Long COVID or post-acute sequelae of COVID-19 (PASC): An overview of biological factors that may contribute to persistent symptoms. Front Microbiol. 2021;12:698169.
16. Roelcke U, Kappos L, Lechner-Scott J, Brunnschweiler H, Huber S, Ammann W, et al. Reduced glucose metabolism in the frontal cortex and basal ganglia of multiple sclerosis patients with fatigue: an 18F-fluorodeoxyglucose positron emission tomography study. Neurology. 1997;48(6):1566–71.
17. Bakshi R, Czarnecki D, Shaikh ZA, Priore RL, Janardhan V, Kaliszky Z, et al. Brain MRI lesions and atrophy are related to depression in multiple sclerosis. Neuroreport. 2000;11(6):1153–8.
18. Blinkenberg M, Rune K, Jensen CV, Ravnborg MH, Kyllingsbaek S, Holm S, et al. Reduced metabolism in cerebral cortex correlates with MRI changes and cognitive dysfunction in patients with disseminated sclerosis. Ugeskr Laeger. 2001;163(27):3788–92.
19. Lulic T, El-Sayes J, Fassett HJ, Nelson AJ. Physical activity levels determine exercise-induced changes in brain excitability. PLoS One. 2017;12(3):e0173672.

20. Campbell M, Varley-Campbell J, Fulford J, Taylor B, Mileva KN, Bowtell JL. Effect of immobilization on neuromuscular function in vivo in humans: a systematic review. Sports Med. 2019;49:931–50.
21. Asadi-Pooya AA, Simani L. Central nervous system manifestations of COVID-19: A systematic review. J Neurol Sci. 2020;413:116832.
22. Morgul E, Bener A, Atak M, Akyel S, Aktas S, Bhugra D, et al. COVID-19 pandemic and psychological fatigue in Turkey. Int J Soc Psychiatry. 2020;1-8:128.
23. Satici B, Gocet-Tekin E, Deniz ME, Satici SA. Adaptation of the fear of COVID-19 scale: its association with psychological distress and life satisfaction in Turkey. Int J Ment Health Addict. 2020; https://doi.org/10.1007/s11469-020-00294-0.
24. Brooks SK, Webster RK, Smith LE, Woodland L, Wessely S, Greenberg N, et al. The psychological impact of quarantine and how to reduce it: rapid review of the evidence. Lancet. 2020;395:912–20.
25. Doykov I, Hällqvist J, Gilmour KC, Grandjean L, Mills K, Heywood WE. The long tail of Covid-19' – The detection of a prolonged inflammatory response after a SARS-CoV-2 infection in asymptomatic and mildly affected patients. F1000Res. 2020;9:1349.
26. Nalbandian A, Sehgal K, Gupta A, Madhavan MV, McGroder C, Stevens JS, et al. Post-acute COVID-19 syndrome. Nat Med. 2021;27(4):601–15.
27. Taylor S. The psychology of pandemics: preparing for the next global outbreak of infectious disease. Cambridge: Cambridge Scholars Publishing; 2019.
28. Disser NP, De Micheli AJ, Schonk MM, Konnaris MA, Piacentini AN, Edon DL, et al. Musculoskeletal Consequences of COVID-19. J Bone Joint Surg Am. 2020;102(14):1197–204.
29. Hoong CWS, Hussain I, Aravamudan VM, Phyu EE, Lin JHX, Koh H. Obesity is associated with poor Covid-19 outcomes: a systematic review and meta-analysis. Horm Metab Res. 2021;53(2):85–93.
30. Chou SH, Beghi E, Helbok R, Moro E, Sampson J, Altamirano V, et al. Global incidence of neurological manifestations among patients hospitalized with COVID-19-A report for the GCS-NeuroCOVID consortium and the ENERGY consortium. JAMA Netw Open. 2021;4(5):e2112131.
31. Mehandru S, Merad M. Pathological sequelae of long-haul COVID. Nat Immunol. 2022;23(2):194–202.
32. Li H, Xue Q, Xu X. Involvement of the nervous system in SARS-CoV-2 infection. Neurotox Res. 2020;38(1):1–7.
33. Bigland-Ritchie B, Rice CL, Garland SJ, Walsh ML. Task-dependent factors in fatigue of human voluntary contractions. Adv Exp Med Biol. 1995;384:361–80.
34. Enoka RM, Stuart DG. Neurobiology of muscle fatigue. J Appl Physiol. 1992;72:1631–48.
35. Enoka RM, Baudry S, Rudroff T, Farina D, Klass M, Duchateau J. Unraveling the neurophysiology of muscle fatigue. J Electromyogr Kinesiol. 2011;21:208–19.
36. Mecenas P, Moreira Bastos RT, Vallinoto AC, Normando D. Effects of temperature and humidity on the spread of COVID-19: a systematic review. PLoS One. 2020;
37. Ssentongo P, Ssentongo AE, Heilbrunn ES, Ba DM, Chinchilli VM. Association of cardiovascular disease and 10 other pre-existing comorbidities with COVID-19 mortality: a systematic review and meta-analysis. PLoS One. 2020;15(8):e0238215.
38. Rudroff T, Workman CD, Ponto LLB. 18F-FDG-PET imaging for post-COVID-19 brain and skeletal muscle alterations. Viruses. 2021;13(11):2283.
39. Rudroff T, Klén R, Rainio O, Tuulari J. The untapped potential of dimension reduction in neuroimaging: artificial intelligence-driven multimodal analysis of long COVID fatigue. Brain Sci. 2024;14:1209.
40. Rudroff T. Artificial intelligence's transformative role in illuminating brain function in long COVID patients using PET/FDG. Brain Sci. 2024;14(1):73.

41. Rudroff T. Frontal-striatal glucose metabolism and fatigue in patients with multiple sclerosis, long COVID, and COVID-19 recovered controls. Exp Brain Res. 2024;242(9):2125–36.
42. Agergaard J, Yamin Ali Khan B, Engell-Sørensen T, Schiøttz-Christensen B, Østergaard L, Hejbøl EK, et al. Myopathy as a cause of Long COVID fatigue: evidence from quantitative and single fiber EMG and muscle histopathology. Clin Neurophysiol. 2023;148:65–75.

4 Sex-Based Differences

4.1 Long COVID Fatigue: Understanding the Female Predominance

Mortality from COVID-19 proves higher in men versus women. Conversely, women demonstrate greater susceptibility to long-lasting impacts like post-COVID fatigue [1–3]. Women 50 years or younger self-report deficient recovery and worse fatigue and disability more frequently than men [4]. Moreover, Sigfrid et al. [4] associated the female sex with elevated pain, anxiety, and depression. The LONG-COVID-Exp-CM Multicenter Study [1] also documented significantly higher post-COVID symptomology (fatigue, dolor, anxiety, depression, poor sleep quality) among women. These findings prompt the question—why do women fare worse? Initial viral exposure differences represent one theory, as occupations like education may heighten women's infection risk [5]. However, data imply teachers lack excess contact versus other working-age groups, alongside indications of sex-based differences in SARS-CoV-2 immune responses [5, 6]. Additionally, given women survive acute SARS-CoV-2 infection more often than men [4], the prevalence of post-COVID fatigue may simply reflect more women progressing past acute infection. Further research should investigate sex-stratified measures of disease severity, considering if women represent a larger pool of survivors in whom to identify symptoms. The precise determinants of post-COVID fatigue among women remain ambiguous. Nevertheless, our prior model [7] spotlights explanatory factors like anxiety, depression, pain, inflammation, and pre-existing illness (Fig. 4.1). Importantly, though, multifaceted origins underline fatigue beyond any individual source.

T. Rudroff, *Long COVID Fatigue*, https://doi.org/10.1007/978-3-031-89474-9_4

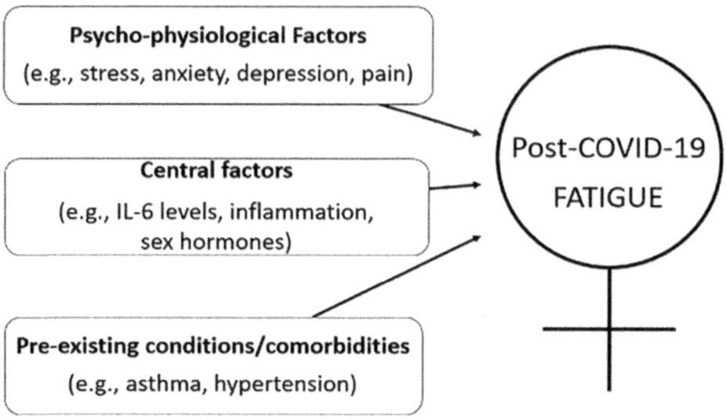

Fig. 4.1 A proposed model for contributors to the increased prevalence of post-COVID-19 fatigue in women. Fatigue is greatly affected by the interactive changes in psycho-physiological factors (stress, anxiety, depression, and pain), central factors (inflammation, sex hormones), and by pre-existing conditions [8]. Note: The term "post-COVID-19 fatigue" used in this published figure is equivalent to "Long COVID fatigue" used throughout this book

4.2 Stress, Anxiety, Depression, and Pain

While quarantines and distancing curb viral transmission, they may adversely intensify fatigue in COVID-19 survivors [9–11]. Concurrent situational depression, anxiety, and stress can also prevail [9]. Women report greater persisting neuropsychiatric symptoms than men [12], suggesting increased susceptibility to enduring COVID-19 neurological and mental health sequelae. Higher self-reported domestic stress among women [12] further implicates home life strain in post-COVID fatigue. Similarly, women demonstrated more psychological effects following the 2003 SARS epidemic [13]. Recent findings also associate absent social support in women with higher COVID-19-related post-traumatic stress [14], indicating greater stress reactivity. Previous cardiovascular research found marital and family stress strongly predicts female atherosclerosis development [15], underscoring the value of managing such stressors [16]. Sex differences in symptom reporting and perception, which could impact interventions and studies, warrant consideration as well.

Biological sex also regulates depressive disorders and COVID-19 outcomes [17]. Women face amplified pandemic-related risks like pre-existing anxiety, depression, chronic stress, and domestic violence [18]. Severe initial COVID-19 commonly precedes longer-term neurological and psychiatric conditions [7, 19], especially among women handling heightened childcare, home education, and family caretaking burdens.

Working mothers may also suffer career and development opportunity costs from increased family obligations. Isolated single women additionally confront pandemic-related economic anxieties and decreased social contact. Subsequent lifestyle changes like unhealthy diets, inactivity, and stress-linked eating/drinking disproportionately predispose women to reliance on food cravings. Such cravings are associated with obesity itself conferring greater COVID-19 risk and inflammation [18]. Besides acute myalgias [20, 21], post-COVID musculoskeletal pain proves highly prevalent months later [1], with de novo symptoms in nearly 75% of those with prior complaints. History of similar pain, extended hospitalizations, headache, and female sex all independently associate with higher risk. Still, women simply perceive more pain than men overall [22].

4.3 Inflammation

Stress, anxiety, depression, pain, and Long COVID fatigue are likely to prove co-occurring and interdependent, sharing inflammation as a common risk factor. Sex differences may regulate profound COVID-19 immune response alterations [23]. Since women disproportionately develop inflammatory disorders exacerbated by stress [24–26], sex-based variances in stress reactions, like interleukin-6 (IL-6) release, could represent a key underlying mechanism [27]. Beyond combatting infection, IL-6 drives inflammation and is associated with fatigue, poor sleep, depression, pain, and affective disturbances [28–30]. In autoimmune and non-autoimmune illnesses alike, IL-6 contributes centrally to fatigue development. Various cell types secrete IL-6 during inflammatory reactions. Accordingly, Durstenfeld et al. [31] found elevated IL-6 among most post-acute COVID-19 sequela patients, proving higher in women and those with fatigue, pain, anxiety, and depression typifying central sensitization. An analogous Mayo Clinic study [32] observed fatigue dominance with the highest IL-6 levels, predominating in women. The majority demonstrated substantial IL-6 elevations up to 3 months post-infection. Moreover, women showed greater stress-induced IL-6 release, implying modulation of their heightened hypothalamic-pituitary-adrenal axis responsiveness [33]. While ambiguous presently, further efforts should clarify IL-6's sex-specific role in post-COVID fatigue, evaluating its usefulness as a virulence biomarker. Research into therapeutic IL-6 blockade may additionally offer insights around persistent viral infection and resultant fatigue.

Sex differences in stress and inflammation regulation likely contribute to COVID-19 dimorphism [34]. Autoimmune conditions like rheumatoid arthritis and lupus preferentially affecting women reinforce this notion [35]. High testosterone exhibits associations with impaired antibody production as well [36]. Compared to men, women's immune systems mount augmented, sustained responses with superior viral clearance [37, 38]. Estrogen modulates inflammation, alongside X chromosome-based immune genes [39], ostensibly attenuating women's cytokine storm severity. Nonetheless, concrete immunological differences in SARS-CoV-2 infection remain unsubstantiated across sexes thus far. Other experts posit overlapping long COVID symptoms with perimenopause and menopause changes highlight the relevance of future sex hormone investigations [40, 41]. Consideration should be given to the higher prevalence of long COVID in women under 50 years old approaching average menopause onset [4, 42].

Summary

The chapter explores the intriguing paradox in COVID-19 outcomes between men and women, where men experience higher mortality rates, but women face greater susceptibility to long-term complications, particularly post-COVID fatigue. The text delves into this gender disparity by examining various contributing factors and potential explanations for women's increased vulnerability to persistent symptoms.

The discussion begins by establishing that women, especially those under 50, report more severe and frequent symptoms of fatigue, disability, pain, anxiety, and depression following COVID-19 infection. While initial theories suggested occupational exposure might explain this disparity, the chapter demonstrates that the relationship is more complex, potentially relating to women's higher survival rates from acute infection and fundamental sex-based differences in immune responses.

The chapter then examines two major mechanistic pathways: psychosocial factors and inflammatory responses. The psychosocial analysis reveals how pandemic-related stressors disproportionately affect women through increased domestic responsibilities, childcare burdens, career impacts, and social isolation. These stressors are compared to historical patterns observed in previous epidemics, particularly SARS, where women also showed greater psychological vulnerability.

The inflammatory component of the discussion is particularly detailed, focusing on the role of interleukin-6 (IL-6) as a key mediator in the relationship between stress, inflammation, and fatigue. The chapter presents evidence showing women heightened inflammatory responses and stress-induced IL-6 release, connecting these biological differences to their increased susceptibility to post-COVID fatigue.

The text concludes by examining the potential role of sex hormones and chromosomal differences in immune responses, suggesting that women's stronger immune responses might paradoxically contribute to their increased risk of long-term symptoms. The chapter also notes the potential overlap between long COVID symptoms and hormonal transitions like perimenopause, highlighting the need for further research into sex-specific aspects of the condition.

References

1. Fernández-De-Las-Peñas C, Martín-Guerrero JD, Pellicer-Valero J, Navarro-Pardo E, Gómez-Mayordomo V, Cuadrado ML, et al. Female sex is a risk factor associated with long-term post-COVID related-symptoms but not with COVID-19 symptoms: the LONG-COVID-EXP-CM multicenter study. J Clin Med. 2022;11:413.
2. Ceban F, Ling S, Lui LM, Lee Y, Gill H, Teopiz KM, et al. Fatigue and cognitive impairment in Post-COVID-19 Syndrome: a systematic review and meta-analysis. Brain Behav Immun. 2022;101:93–135.

3. Bechmann N, Barthel A, Schedl A, Herzig S, Varga Z, Gebhard C. Sexual dimorphism in COVID-19: potential clinical and public health implications. Lancet Diabetes Endocrinol. 2022;10:221–30.
4. Sigfrid L, Drake TM, Pauley E, Jesudason EC, Olliaro O, Lim WS, et al. Long Covid in adults discharged from UK hospitals after Covid-19: a prospective, multicentre cohort study using the ISARIC WHO Clinical Characterization Protocol. Lancet Reg Health Eur. 2021;8:100186.
5. Office for National Statistics. Coronavirus (COVID-19), Latest data and analysis on coronavirus (COVID-19) in the UK and its effect on the economy and society. London: ONS; 2025.
6. Takahashi T, Ellingson MK, Wong P, Israelow B, Lucas C, Klein J, et al. Sex differences in immune responses that underlie COVID-19 disease outcomes. Nature. 2020;588:315–20.
7. Rudroff T, Fietsam AC, Deters JR, Bryant AD, Kamholz J. Post-COVID-19 fatigue: potential contributing factors. Brain Sci. 2020;10:1012.
8. Rudroff T, Workman CD, Bryant AD. Potential factors that contribute to Post-COVID-19 fatigue in women. Brain Sci. 2022;12:556.
9. Morgul E, Bener A, Atak M, Akyel S, Aktas S, Bhugra D, et al. COVID-19 pandemic and psychological fatigue in Turkey. Int J Soc Psychiatry 2020:1–8.
10. Satici B, Gocet-Tekin E, Deniz ME, Satici SA. Adaptation of the fear of COVID-19 scale: its association with psychological distress and life satisfaction in Turkey. Int J Ment Health Addict. 2020;
11. Brooks SK, Webster RK, Smith LE, Woodland L, Wessely S, Greenberg N, et al. The psychological impact of quarantine and how to reduce it: rapid review of the evidence. Lancet. 2020;395:912–20.
12. Gebhard CE, Sütsch C, Gebert P, Gysi B, Bengs S, Todorov A, et al. Impact of sex and gender on post-COVID-19 syndrome, Switzerland, 2020. Euro Surveill. 2024;29:2300200.
13. Lee AM, Wong JG, McAlonan GM, Cheung V, Cheung C, Sham PC, et al. Stress and psychological distress among SARS survivors 1 year after the outbreak. Can J Psychiat. 2007;52:233–40.
14. Cai Q, Feng H, Huang J, Wang M, Wang Q, Lu X, et al. The mental health of frontline and non-frontline medical workers during the coronavirus disease 2019 (COVID-19) outbreak in China: a case-control study. J Affect Disord. 2020;275:210–5.
15. Wang PS, Aguilar-Gaxiola S, Alonso J, Angermeyer MC, Borges G, Bromet EJ, et al. Use of mental health services for anxiety, mood, and substance disorders in 17 countries in the WHO world mental health surveys. Lancet. 2007;370:841–50.
16. Orth-Gomér K, Schneiderman N, Wang HX, Walldin C, Blom M, Jernberg T. Stress reduction prolongs life in women with coronary disease: the Stockholm Women's Intervention Trial for Coronary Heart Disease. Circ Cardiovasc Qual Outcomes. 2009;2:25–32.
17. Almeida M, Shrestha AD, Stojanac D, Miller LJ. The impact of the COVID-19 pandemic on women's mental health. Arch Womens Ment Health. 2020;23:741–8.
18. Mattioli AV, Pinti M, Farinetti A, Nasi M. Obesity risk during collective quarantine for the COVID-19 epidemic. Obes Med. 2020;20:100263.
19. Taquet M, Geddes JR, Husain M, Luciano S, Harrison PJ. 6-month neurological and psychiatric outcomes in 236 379 survivors of COVID-19: a retrospective cohort study using electronic health records. Lancet Psychiatry. 2021;8:416–27.
20. Abdullahi A, Candan SA, Abba MA, Bello AH, Alshehri MA, Victor EA, et al. Neurological and musculoskeletal features of COVID-19: a systematic review and meta-analysis. Front Neurol. 2020;11:687.
21. Ciaffi J, Meliconi R, Ruscitti P, Berardicurti O, Giacomelli R, Ursini F. Rheumatic manifestations of COVID-19: a systematic review and metaanalysis. BMC Rheumatol. 2020;4:65.
22. Barsky AJ, Peekna HM, Borus JF. Somatic symptom in women and men. J Gen Intern Med. 2001;16:266–75.
23. Rabaan AA, Al-Ahmed SH, Garout MA, Al-Quaaneh AM, Sule AA, Tirupathi R, et al. Diverse immunological factors influencing pathogenesis in patients with COVID-19: a review on viral Dissemination, Immunotherapeutic options to counter cytokine storm and inflammatory responses. Pathogens. 2021;10:565.

24. Whitacre CC, Reingold SC, O'Looney PA. A gender gap in autoimmunity. Science. 1999;283:1277–8.
25. Kimyai-Asadi A, Usman A. The role of psychological stress in skin disease. J Cutan Med Surg. 2001;5:140–5.
26. Lechin F, van der Dijs B, Jackubowicz D, Camero RE, Lechin S, Villa S, et al. Role of stress in the exacerbation of chronic illness. Psychoneuroendocrinology. 1987;12:117–29.
27. Edwards KM, Burns VE, Ring C, Carroll D. Sex difference in the interleukin-6 response to acute psychological stress. Biol Psychol. 2006;71:236–9.
28. Rohleder N, Aringer M, Boentert M. Role of interleukin-6 in stress, sleep, and fatigue. Ann N Y Acad Sci. 2012;1261:88–96.
29. Grygiel-Górniak B, Puszczewicz M. Fatigue and interleukin-6—a multi-faceted relationship. Reumatologia. 2015;53:207–12.
30. Bossola M, Di Stasio E, Giungi S, Rosa F, Tazza L. Fatigue is associated with serum interleukin-6 levels and symptoms of depression in patients on chronic hemodialysis. J Pain Symptom Manage. 2015;49:578–85.
31. Durstenfeld MS, Peluso MJ, Kelly JD, Win S, Swaminathan S, Li D, et al. Role of antibodies, inflammatory markers, and echocardiographic findings in post-acute cardiopulmonary symptoms after SARS-CoV-2 infection. medRxiv. 2021;
32. Ganesh R, Grach SL, Ghosh AK, Bierle DM, Salonen BR, Collins NM, et al. The female-predominant persistent immune dysregulation of the post-COVID syndrome. Mayo Clin Proc. 2022;97:454–64.
33. Jankord R, Turk JR, Schadt JC, Casati J, Ganjam VK, Price EM, et al. Sex difference between interleukin-6 and stress. Endocrinology. 2007;148:3758–64.
34. Veldhuis JD, Sharma A, Roelfsema F. Age-dependent and gender dependent regulation of hypothalamic-adrenocorticotropic-adrenal axis. Endocrinol Metab Clin N Am. 2013;42:201–25.
35. Amur S, Parekh A, Mummaneni P. Sex differences and genomics in autoimmune diseases. J Autoimmun. 2012;38:J254–65.
36. Acharya Y, Pant S, Gyanwali P, Dangal G, Karki P, Bista NR, et al. Gender disaggregation in COVID-19 and increased male susceptibility. J Nepal Health Res Counc. 2020;18:345–50.
37. Falahi S, Kenarkoohi A. Sex and differences in the outcome of patients with COVID-19. J Med Virol. 2021;93:151–2.
38. Jin JM, Bai P, He W, Wu F, Liu XF, Han DM, et al. Gender differences in patients with COVID-19: Focus on severity and mortality. Front Public Health. 2020;8:152.
39. Conti P, Younes A. Coronavirus COV-19/SARS-CoV-2 affects women less than men: clinical response to viral infection. J Biol Regul Homeost Agents. 2020;34:2.
40. Stewart S, Newson L, Briggs TA, Grammatopoulos D, Young L, Gill P. Long COVID Risk—A signal to address sex hormones and women's health. Lancet Reg Health Eur. 2021;11:100242.
41. Sudre CH, Murray B, Varsavsky T, Graham MS, Penfold RS, Bowyer RC, et al. Attributes and predictors of long COVID. Nat Med. 2021;27:626–31.
42. National Institute for Health and Care Excellence. COVID-19 rapid guideline: managing the long-term effects of COVID-19. London: NICE; 2022.

Age-Related Considerations 5

5.1 Age-Related Impact of Long COVID Fatigue

The impact of Long COVID fatigue varies significantly across age groups, with distinct manifestations, prevalence rates, and management considerations for different populations (Fig. 5.1). Understanding these age-related differences is crucial for appropriate clinical assessment and intervention planning.

The manifestation and impact of Long COVID fatigue demonstrate significant age-dependent variations, encompassing differences in prevalence, clinical presentation, and management requirements across the lifespan. The PHOSP-COVID Collaborative Group [1] has documented a clear age-dependent increase in prevalence, ranging from 5% to 7% in pediatric populations to 20–25% in elderly individuals, highlighting the increased vulnerability with advancing age.

In the pediatric population (0–18 years), Long COVID fatigue presents unique challenges related to developmental trajectories and educational impact. Buonsenso et al. [2], in their comprehensive study published in The Lancet Child & Adolescent Health, identified distinctive manifestations including mood changes, sleep disturbances, and academic difficulties. These symptoms require specific consideration of developmental impacts and educational accommodations, with particular attention to family support systems and long-term monitoring of developmental progression.

Young adults (19–40 years) experience a different spectrum of challenges, with prevalence rates of 10–15% reported in longitudinal studies [1]. This age group predominantly reports physical fatigue, significant cognitive dysfunction ("brain fog"), and substantial impact on work capabilities. The comprehensive analysis by Fernández-de-las-Peñas et al. [3] emphasized the particular challenge of career disruption and the need for workplace accommodations in this demographically active workforce population.

Middle-aged adults (41–65 years) show increased prevalence (15–20%) and complexity of presentation. The PHOSP-COVID Collaborative Group [1]

T. Rudroff, *Long COVID Fatigue*, https://doi.org/10.1007/978-3-031-89474-9_5

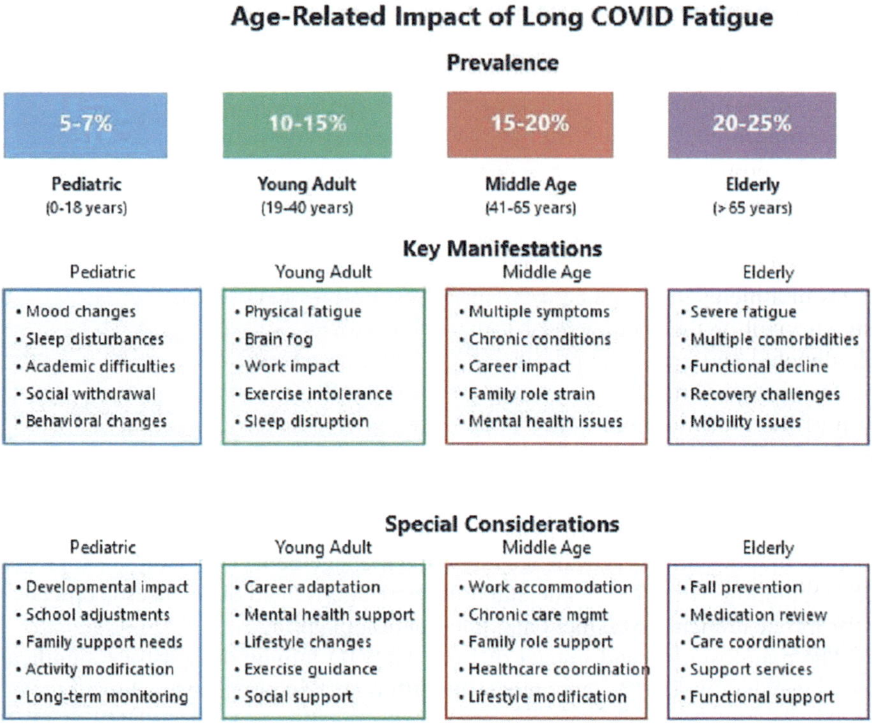

Fig. 5.1 Age-related impact of Long COVID fatigue: Prevalence, manifestations, and special considerations

documented multiple concurrent symptoms, often complicated by pre-existing conditions and significant impacts on both professional and family roles. This age group frequently requires comprehensive management strategies addressing both physical symptoms and psychosocial support.

Using Medicare data and a symptom-based definition, Fung et al. [4] estimate that long COVID happens in 16.6% of outpatients and 29.2% of hospitalized patients that are over 65. Walle-Hansen et al. [5] report significant deterioration in quality of life and functional decline in elderly patients 6 months after COVID-19. More information is needed to understand the impact of long COVID on the elderly population.

Special considerations vary significantly across age groups. Pediatric populations require careful attention to developmental impacts and educational support [2], while young adults need focused support for career adaptation and lifestyle modifications [1]. Middle-aged adults often require complex care coordination addressing both professional and family responsibilities [1], and elderly individuals need comprehensive geriatric assessment and support services.

The subsequent chapters will provide detailed examinations of Long COVID fatigue in elderly populations and children, offering in-depth analysis of the unique challenges and management strategies for these particularly vulnerable age groups. These focused discussions will expand upon the age-specific considerations introduced here, providing healthcare providers with detailed guidance for these distinct populations.

5.2 Aging and Susceptibility to Long COVID Fatigue

Older adults are at an increased risk of severe COVID-19 illness and mortality, primarily due to age-related changes in the immune system, the presence of comorbidities, and the effects of immunosenescence [6]. Growing evidence suggests that older age is also a significant risk factor for developing Long COVID and its associated symptoms, including fatigue [7].

A large-scale study by Sudre et al. [7] using data from the COVID Symptom Study app found that older age was independently associated with an increased likelihood of experiencing persistent fatigue following COVID-19 infection. Individuals over the age of 50 were more likely to report fatigue lasting beyond 28 days compared to younger adults [7].

Similarly, a systematic review and meta-analysis by Michelen et al. [8] found that increasing age was a significant risk factor for developing post-acute COVID-19 syndrome, with fatigue being one of the most common symptoms reported. The study highlighted the need for long-term monitoring and support for older adults recovering from COVID-19 [8].

The mechanisms underlying the increased susceptibility of older adults to Long COVID fatigue are likely multifactorial and may involve a combination of immunological, neurological, and metabolic factors [9].

5.3 Immunosenescence and Inflammation

Immunosenescence, the age-related decline in immune function, may play a crucial role in the development of Long COVID fatigue in older adults [10]. With advancing age, the immune system undergoes a series of changes, including a reduction in the number and functionality of immune cells, altered cytokine production, and a chronic low-grade inflammatory state known as inflammaging [11].

In the context of COVID-19, immunosenescence may lead to a less effective initial immune response to the virus, resulting in prolonged viral persistence and a heightened inflammatory response [12]. This persistent inflammation, characterized by elevated levels of pro-inflammatory cytokines such as interleukin-6 (IL-6) and tumor necrosis factor-alpha (TNF-α), has been implicated in the development of fatigue and other Long COVID symptoms [13].

A study by Peluso et al. [14] found that older adults with Long COVID had significantly higher levels of inflammatory markers, including IL-6 and C-reactive protein (CRP), compared to younger adults with Long COVID and healthy controls. The authors suggested that the persistent inflammation observed in older adults may contribute to the development and maintenance of fatigue and other Long COVID symptoms [14].

5.4 Neurological Implications

Aging is associated with a variety of neurological changes that may increase the vulnerability of older adults to the neurological manifestations of Long COVID, including fatigue [15]. Age-related alterations in the central nervous system, such as reduced neurotransmitter levels, impaired mitochondrial function, and increased oxidative stress, may contribute to the development of fatigue and cognitive dysfunction in older adults with Long COVID [16].

A study by Hosp et al. [17] using positron emission tomography (PET) imaging found that older adults with Long COVID exhibited reduced glucose metabolism in the frontal cortex and basal ganglia, which was associated with fatigue severity. The authors hypothesized that the neuroinflammation and metabolic dysregulation observed in these brain regions may underlie the persistent fatigue experienced by older adults with Long COVID [17].

Furthermore, age-related changes in the autonomic nervous system, such as reduced heart rate variability and impaired baroreflex sensitivity, may contribute to the development of orthostatic intolerance and post-exertional malaise, which are common features of Long COVID fatigue [18].

5.5 Comorbidities and Frailty

The presence of pre-existing comorbidities and frailty in older adults may further complicate the development and management of Long COVID fatigue [19]. Conditions such as cardiovascular disease, diabetes, and chronic obstructive pulmonary disease (COPD) are more prevalent in older adults and have been associated with an increased risk of severe COVID-19 illness and persistent symptoms, including fatigue [20].

Frailty, a state of increased vulnerability to stressors due to age-related declines in physiological reserve and function, has also been identified as a significant risk factor for adverse outcomes in older adults with COVID-19 [21]. Frail older adults may experience a more prolonged and challenging recovery process, with a higher likelihood of developing Long COVID fatigue and other persistent symptoms [4, 5].

5.6 Management Strategies

Managing Long COVID fatigue in older adults requires a multidisciplinary and personalized approach that considers the unique challenges and needs of this population [22]. A combination of pharmacological and non-pharmacological interventions may be necessary to address the complex interplay of factors contributing to fatigue in older adults with Long COVID.

Non-pharmacological interventions, such as pacing, energy conservation techniques, and graded exercise therapy, have shown promise in managing fatigue in older adults with Long COVID [23]. These strategies aim to help patients balance rest and activity, minimize energy expenditure, and gradually increase exercise tolerance while avoiding post-exertional malaise.

Cognitive behavioral therapy (CBT) and other psychological interventions may also be beneficial in addressing the emotional and cognitive aspects of Long COVID fatigue in older adults. CBT can help patients develop coping strategies, manage stress and anxiety, and improve sleep quality, all of which may contribute to a reduction in fatigue severity [24].

Pharmacological treatments for Long COVID fatigue in older adults may include medications targeting specific symptoms or underlying mechanisms, such as stimulants, anti-inflammatory agents, or neurotransmitter modulators [25]. However, the use of pharmacological interventions in this population should be carefully considered and monitored, given the increased risk of adverse effects and drug interactions in older adults.

5.7 Future Directions and Research

Priorities as the long-term consequences of COVID-19 continue to unfold, further research is needed to better understand the unique challenges and needs of older adults with Long COVID fatigue. Future studies should focus on:

1. Investigating the specific biological mechanisms underlying the increased susceptibility of older adults to Long COVID fatigue, including immunosenescence, neuroinflammation, and metabolic dysregulation [12].
2. Developing and validating age-specific screening tools and diagnostic criteria for Long COVID fatigue to facilitate early identification and intervention in older adults [26].
3. Conducting randomized controlled trials to evaluate the safety and efficacy of pharmacological and non-pharmacological interventions for Long COVID fatigue in older adults, considering the potential for age-related differences in treatment response and tolerability [25].
4. Exploring the role of personalized medicine approaches, such as genetic profiling and biomarker-guided treatment, in optimizing the management of Long COVID fatigue in older adults [27].
5. Investigating the long-term impact of Long COVID fatigue on functional outcomes, quality of life, and healthcare utilization in older adults, and developing strategies to mitigate these consequences [4].

Summary

In this chapter presents a comprehensive examination of how long COVID fatigue manifests and impacts individuals across different age groups, revealing a clear pattern of increasing prevalence with age. The research shows a striking progression, beginning with relatively low rates of 5–7% in pediatric populations and climbing significantly to reach nearly 30% among hospitalized elderly patients.

The chapter details how each age group faces distinct challenges. Children and adolescents must navigate the complex interplay between their developmental trajectory and the impact of long COVID fatigue on their educational journey, often experiencing mood fluctuations and disrupted sleep patterns. Young adults, in their prime working years, grapple with the dual burden of physical exhaustion and cognitive dysfunction, commonly described as "brain fog," which can significantly derail their career progression.

Middle-aged adults present a more intricate clinical picture, with long COVID fatigue often interweaving with pre-existing conditions, creating a complex web of symptoms that affect both their professional lives and family responsibilities. The prevalence in this group ranges from 15% to 20%, highlighting the significant burden on this demographically active population.

The elderly population emerges as particularly vulnerable, with the chapter dedicating substantial attention to understanding why older adults face heightened susceptibility. The text explores several key factors, including immunosenescence, which compromises the aging immune system's ability to respond effectively to the virus. This biological change, combined with age-related neurological alterations and the presence of pre-existing conditions, creates a perfect storm that increases both the likelihood and severity of long COVID fatigue in older adults.

Management strategies are presented through a nuanced, age-sensitive lens. The chapter advocates for a comprehensive approach that combines various interventions, from carefully paced physical activity to psychological support through cognitive behavioral therapy. For older adults, particular attention is paid to the delicate balance required in pharmacological interventions, acknowledging the increased risks of adverse effects and drug interactions in this population.

The chapter concludes by charting a course for future research, emphasizing the need to better understand the biological mechanisms underlying age-related susceptibility, develop more precise diagnostic tools, and evaluate the long-term implications for quality of life and healthcare utilization. This forward-looking perspective underscores the dynamic nature of our understanding of long COVID fatigue and the continuing need for age-specific research and interventions.

Throughout, the chapter maintains a clear focus on the practical implications for healthcare providers, offering guidance for tailoring approaches to each age group while acknowledging the unique challenges and needs of different populations. This comprehensive understanding of age-related variations in long COVID fatigue is presented as crucial for developing effective, personalized treatment strategies.

References

1. PHOSP-COVID Collaborative Group. Clinical characteristics with inflammation profiling of long COVID and association with 1-year recovery following hospitalization in the UK: a prospective observational study. Lancet Respir Med. 2022;10(8):761–75.
2. Buonsenso D, Espuny Pujol F, Munblit D, Pata D, McFarland S, Simpson FK. Clinical characteristics, activity levels and mental health problems in children with long coronavirus disease: a survey of 510 children. Future Microbiol. 2022;17(8):577–88.
3. Fernández-de-las-Peñas C, de-la-Llave-Rincon AI, Ortega-Santiago R, Ambite-Quesada S, Gomez-Mayordormo V, Cuadrado ML, et al. Prevalence and risk factors of musculoskeletal pain symptoms as long-term post-COVID sequelae in hospitalized COVID-19 survivors: a multicenter study. Pain. 2022;163:1–8.
4. Fung KW, Baye F, Baik SH, Zheng Z, McDonald CJ. Prevalence and characteristics of long COVID in elderly patients: an observational cohort study of over 2 million adults in the US. PLoS Med. 2023;20(4):e1004194.
5. Walle-Hansen MM, Ranhoff AH, Mellingsaeter M, Wang-Hansen MS, Myrstad M. Health-related quality of life, functional decline, and long-term mortality in older patients following hospitalization due to COVID-19. BMC Geriatr. 2021;21(1):199.
6. Mueller AL, McNamara MS, Sinclair DA. Why does COVID-19 disproportionately affect older people? Aging (Albany NY). 2020;12(10):9959–81.
7. Sudre CH, Murray B, Varsavsky T, Graham MS, Penfold RS, Bowyer RC, et al. Attributes and predictors of long COVID. Nat Med. 2021;27(4):626–31.
8. Michelen M, Manoharan L, Elkheir N, Cheng V, Dagens A, Hastie C, et al. Characterizing long COVID: a living systematic review. BMJ Glob Health. 2021;6(9):e005427.
9. Akbar AN, Gilroy DW. Aging immunity may exacerbate COVID-19. Science. 2020;369(6501):256–7.
10. Hazeldine J, Lord JM. Immunesenescence: a predisposing risk factor for the development of COVID-19? Front Immunol. 2020;11:573662.
11. Franceschi C, Garagnani P, Parini P, Giuliani C, Santoro A. Inflammaging: a new immune-metabolic viewpoint for age-related diseases. Nat Rev Endocrinol. 2018;14(10):576–90.
12. Bektas A, Schurman SH, Sen R, Ferrucci L. Aging, inflammation and the environment. Exp Gerontol. 2020;130:110793.
13. Doykov I, Hällqvist J, Gilmour KC, Grandjean L, Mills K, Heywood WE. 'The long tail of Covid-19' – the detection of a prolonged inflammatory response after a SARS-CoV-2 infection in asymptomatic and mildly affected patients. F1000Research. 2021;9:1349.
14. Peluso MJ, Lu S, Tang AF, Durstenfeld MS, Ho HE, Goldberg SA, et al. Markers of immune activation and inflammation in individuals with Postacute sequelae of severe acute respiratory syndrome coronavirus 2 infection. J Infect Dis. 2021;224(11):1839–48.
15. Jaywant A, Vanderlind WM, Alexopoulos GS, Fridman CB, Perlis RH, Gunning FM. Frequency and profile of objective cognitive deficits in hospitalized patients recovering from COVID-19. Neuropsychopharmacology. 2021;46:2235–40.
16. Zampino M, Spencer RG, Fishbein KW, Simonsick EM, Ferrucci L. Cardiovascular health and mitochondrial function: testing an association. J Gerontol A Biol Sci Med Sci. 2021;76(2):361–7.
17. Hosp JA, Dressing A, Blazhenets G, Bormann T, Rau A, Schwabenland M, et al. Cognitive impairment and altered cerebral glucose metabolism in the subacute stage of COVID-19. Brain. 2021;144(4):1263–76.
18. Goldstein DS. The possible association between COVID-19 and postural tachycardia syndrome. Heart Rhythm. 2021;18(4):508–9.
19. Landi F, Barillaro C, Bellieni A, Brandi V, Carfì A, Cipriani MC, et al. The new challenge of geriatrics: saving frail older people from the SARS-CoV-2 pandemic infection. J Nutr Health Aging. 2021;24(5):466–70.

20. Bajaj V, Gadi N, Spihlman AP, Wu SC, Choi CH, Moulton VR. Aging, immunity, and COVID-19: how age influences the host immune response to coronavirus infections. Front Physiol. 2021;11:571416.
21. Hewitt J, Carter B, Vilches-Moraga A, Quinn TJ, Braude P, Verduri A, et al. The effect of frailty on survival in patients with COVID-19 (COPE): a multicentre, European, observational cohort study. Lancet Public Health. 2020;5(8):e444–51.
22. Greenhalgh T, Knight M, A'Court C, Buxton M, Husain L. Management of post-acute covid-19 in primary care. BMJ. 2020;370:m3026.
23. Décary S, Gaboury I, Poirier S, Garcia C, Simpson S, Bull M, et al. Humility and acceptance: working within our limits with long COVID and Myalgic encephalomyelitis/chronic fatigue syndrome. J Orthop Sports Phys Ther. 2021;51(5):197–200.
24. Kuut TA, Müller F, Csorba I, Braamse A, Aldenkamp A, Appelman B, et al. Efficacy of cognitive-behavioral therapy targeting severe fatigue following coronavirus disease 2019: results of a randomized controlled trial. Clin Infect Dis. 2023;77(5):687–95.
25. O'Brien KK, Ford HL, Fayed N, Solomon P. Long COVID and episodic disability: advancing the conceptualization, measurement and knowledge of episodic disability among people living with Long COVID. BMJ Open. 2021;11(8):e049805.
26. Parkin A, Davison J, Tarrant R, Ross D, Halpin S, Simms A, et al. A multidisciplinary NHS COVID-19 service to manage post-COVID-19 syndrome in the community. J Prim Care Community Health. 2021;12:21501327211010994.
27. Compagno S, Palermi S, Pescatore V, Brugin E, Sarto M, Marin R, et al. Physical and psychological reconditioning in long COVID syndrome: results of an out-of-hospital exercise and psychological – based rehabilitation program. Int J Cardiol Heart Vasc. 2022;41:101080.

Children and Long COVID Fatigue

6

6.1 Prevalence and Characteristics

The prevalence of Long COVID in children and adolescents is still being investigated, with estimates varying widely across studies due to differences in case definitions, study populations, and follow-up periods [1]. A systematic review by Lopez-Leon et al. [2] found that the prevalence of persistent symptoms in children and adolescents following SARS-CoV-2 infection ranged from 4% to 66%, with fatigue being one of the most frequently reported symptoms.

A large-scale study by Stephenson et al. [3] investigated the long-term health outcomes of SARS-CoV-2 infection in children and young people aged 11–17 years in the United Kingdom. The study found that 66.5% of participants reported at least one persistent symptom 15 weeks after testing positive for SARS-CoV-2, with fatigue being reported by 39.0% of participants [3].

Another study by Buonsenso et al. [4] assessed the persistence of symptoms in children and adolescents aged 0–18 years who had been diagnosed with COVID-19 in Italy. The study found that 53.3% of participants reported at least one persistent symptom 120 days after diagnosis, with fatigue being reported by 10.8% of participants [4].

A recent comparative study by Conde et al. [5] provides additional insights into the persistence of symptoms among hospitalized children. Their research, conducted across three hospitals in Madrid, examined 96 children—50 with acute COVID-19 and 46 hospitalized for other conditions. While they found that persistent symptoms were more common in COVID-19 patients (28% vs. 15% at 1 year), this difference did not reach statistical significance. The symptom patterns differed notably between groups: COVID-19 patients predominantly experienced fatigue (8%) and headaches (6%), whereas non-COVID-19 patients more commonly reported abdominal pain and poor appetite (7% each). Interestingly, the study revealed that children who experienced COVID-19 reinfections were less likely to develop persistent symptoms compared to those with single infections. This research

T. Rudroff, *Long COVID Fatigue*, https://doi.org/10.1007/978-3-031-89474-9_6

suggests that while Long COVID symptoms may be more prevalent in hospitalized children with COVID-19, persistent post-hospitalization symptoms are not unique to COVID-19 infection, highlighting the complexity of post-acute sequelae in pediatric populations.

This evidence adds an important dimension to our understanding of Long COVID fatigue in children, particularly in the context of hospitalization. It emphasizes the need to consider both COVID-specific and general post-hospitalization factors when evaluating persistent symptoms in pediatric populations. The study's findings of distinct symptom patterns between COVID-19 and non-COVID-19 patients may also help clinicians better differentiate between post-COVID sequelae and other post-hospitalization conditions.

The characteristics of Long COVID fatigue in children and adolescents may differ from those observed in adults. Children and adolescents with Long COVID fatigue often report a fluctuating pattern of symptoms, with periods of relative well-being interspersed with episodes of severe fatigue and other symptoms [6]. The impact of fatigue on daily functioning may also be more pronounced in children and adolescents, as it can interfere with their ability to attend school, participate in extracurricular activities, and maintain social connections [7].

6.2 Potential Mechanisms

The mechanisms underlying Long COVID fatigue in children and adolescents are not yet fully understood, but several hypotheses have been proposed based on the current evidence and knowledge of other post-viral fatigue syndromes, such as chronic fatigue syndrome/myalgic encephalomyelitis (CFS/ME) [6].

One potential mechanism is persistent immune dysregulation and inflammation following SARS-CoV-2 infection [6]. Children and adolescents with Long COVID fatigue have been found to exhibit elevated levels of pro-inflammatory cytokines, such as interleukin-6 (IL-6) and tumor necrosis factor-alpha (TNF-α), which have been implicated in the development of fatigue in other chronic conditions [8]. The persistent inflammatory state may contribute to the ongoing symptoms and disability observed in children and adolescents with Long COVID fatigue [6].

Another potential mechanism is autonomic dysfunction, which has been reported in children and adolescents with Long COVID fatigue [9]. Autonomic dysfunction can manifest as symptoms such as orthostatic intolerance, postural orthostatic tachycardia syndrome (POTS), and impaired heart rate variability, which may contribute to the experience of fatigue and other symptoms [9]. The mechanisms underlying autonomic dysfunction in Long COVID are not yet clear but may involve persistent inflammation, autoimmunity, or direct viral effects on the autonomic nervous system [6].

Psychosocial factors, such as stress, anxiety, and social isolation, may also play a role in the development and maintenance of Long COVID fatigue in children and adolescents [6]. The pandemic has had a significant impact on the mental health and well-being of children and adolescents, and those with Long COVID may be particularly vulnerable to the psychosocial consequences of the pandemic [10]. Addressing these psychosocial factors may be an important component of the management of Long COVID fatigue in children and adolescents [5].

6.3 Management Strategies

The management of Long COVID fatigue in children and adolescents requires a multidisciplinary and individualized approach that addresses the complex interplay of biological, psychological, and social factors contributing to the condition [6]. The following strategies have been proposed for the management of Long COVID fatigue in children and adolescents:

1. Pacing and energy conservation: Pacing involves carefully balancing periods of activity with periods of rest to avoid overexertion and prevent the worsening of symptoms [11]. Children and adolescents with Long COVID fatigue should be encouraged to prioritize essential activities, break tasks into smaller, manageable steps, and take regular breaks throughout the day [6]. Energy conservation techniques, such as using assistive devices or modifying the environment, can also help reduce the energy demands of daily activities [11].
2. Graded exercise therapy (GET): GET is a structured exercise program that aims to gradually increase the duration and intensity of physical activity over time, with the goal of improving physical functioning and reducing fatigue [11]. However, the use of GET in children and adolescents with Long COVID fatigue remains controversial, as some patients may experience post-exertional malaise (PEM) and a worsening of symptoms following exercise [6]. A cautious and individualized approach to physical activity, guided by the patient's symptoms and tolerability, is recommended [6].
3. Cognitive-behavioral therapy (CBT): CBT is a psychotherapeutic approach that aims to help patients identify and modify maladaptive thoughts and behaviors that contribute to fatigue and other symptoms [10]. CBT has been shown to be effective in managing fatigue and improving quality of life in children and adolescents with other chronic conditions, such as CFS/ME [10]. CBT may be particularly useful in addressing the psychosocial factors contributing to Long COVID fatigue in children and adolescents, such as stress, anxiety, and social isolation [6].
4. Pharmacological interventions: There is currently limited evidence for the use of pharmacological interventions in the management of Long COVID fatigue in children and adolescents [11]. However, some medications, such as low-dose naltrexone, have shown promise in the treatment of other chronic fatigue syndromes and may be considered on a case-by-case basis [11]. Other medications, such as those targeting specific symptoms like pain or sleep disturbances, may also be used as part of a comprehensive management plan [6].

5. Education and support: Education and support are essential components of the management of Long COVID fatigue in children and adolescents [6]. Children and their families should be provided with clear and accurate information about the condition, its potential impact on daily life, and the available management strategies [7]. Support groups and online resources can also be valuable sources of information and emotional support for children and families affected by Long COVID fatigue [7].

6.4 Future Directions and Research Priorities

As the understanding of Long COVID fatigue in children and adolescents continues to evolve, several key areas for future research have been identified:

1. Epidemiological studies: Large-scale, prospective studies are needed to better understand the prevalence, risk factors, and long-term outcomes of Long COVID fatigue in children and adolescents [12]. These studies should use standardized case definitions and outcome measures to facilitate comparisons across different populations and settings [1].
2. Mechanistic studies: Further research is needed to elucidate the biological mechanisms underlying Long COVID fatigue in children and adolescents, including the role of persistent inflammation, autonomic dysfunction, and mitochondrial dysfunction [6]. These studies may involve the use of advanced imaging techniques, such as functional magnetic resonance imaging (fMRI) and positron emission tomography (PET), as well as the assessment of biomarkers of immune activation and energy metabolism.
3. Intervention studies: Randomized controlled trials are needed to evaluate the safety and efficacy of different management strategies for Long COVID fatigue in children and adolescents, including pacing, graded exercise therapy, cognitive-behavioral therapy, and pharmacological interventions [11]. These studies should also assess the long-term outcomes and potential adverse effects of these interventions [6].
4. Qualitative studies: Qualitative research methods, such as interviews and focus groups, can provide valuable insights into the lived experiences of children and adolescents with Long COVID fatigue, as well as their families and caregivers [10]. These studies can help identify the specific challenges and unmet needs of this population, as well as the factors that facilitate or hinder their recovery and well-being [7].
5. Collaborative research: Given the complexity and multidisciplinary nature of Long COVID fatigue in children and adolescents, collaborative research efforts involving clinicians, researchers, and patient advocates from different fields and settings are essential [12]. These efforts can help foster the exchange of knowledge and best practices, as well as the development of evidence-based guidelines and recommendations for the management of Long COVID fatigue in children and adolescents [1].

Summary

This comprehensive chapter explores Long COVID fatigue in pediatric populations, beginning with current epidemiological data that show persistent symptom rates ranging from 4% to 66% in children and adolescents. The chapter emphasizes the variability in prevalence across different studies, highlighting recent research, including a comparative study from Madrid that found higher rates of persistent symptoms in COVID-19 patients compared to those hospitalized for other conditions, though the difference wasn't statistically significant. The text delves into the unique characteristics of Long COVID fatigue in children, noting how symptoms often fluctuate and can significantly impact educational and social development. The chapter explores various potential underlying mechanisms, including immune dysregulation, autonomic dysfunction, and psychosocial factors, drawing parallels with other post-viral fatigue syndromes.

A significant portion of the chapter is dedicated to management strategies, advocating for a multidisciplinary approach. These strategies include carefully structured pacing and energy conservation techniques, cautious implementation of graded exercise therapy, cognitive-behavioral therapy, and selective use of pharmacological interventions when appropriate. The text emphasizes the importance of education and support for both children and their families throughout the management process.

The chapter concludes by outlining critical areas for future research, calling for more extensive epidemiological studies, deeper investigation into biological mechanisms, and evaluation of various intervention strategies. It emphasizes the need for standardized case definitions and outcome measures to facilitate meaningful comparison across studies. The text particularly highlights the importance of collaborative research efforts and qualitative studies to better understand the lived experiences of affected children and their families.

Throughout, the chapter maintains a balanced perspective, acknowledging both the complexity of the condition and the current limitations in our understanding while providing practical guidance for clinical management and future research directions.

References

1. Radtke T, Ulyte A, Puhan MA, Kriemler S. Long-term symptoms after SARS-CoV-2 infection in children and adolescents. JAMA. 2021;326(9):869–71.
2. Lopez-Leon S, Wegman-Ostrosky T, Perelman C, Sepulveda R, Rebolledo PA, Cuapio A, et al. More than 50 long-term effects of COVID-19: a systematic review and meta-analysis. Sci Rep. 2021;11(1):16144.
3. Stephenson T, Pinto Pereira SM, Shafran R, de Stavola BL, Rojas N, McOwat K, et al. Physical and mental health 3 months after SARS-CoV-2 infection (long COVID) among adolescents in England: a national matched cohort study. Lancet Child Adolesc Health. 2022;6(4):230–9.

4. Buonsenso D, Espuny Pujol F, Munblit D, Pata D, McFarland S, Simpson FK. Clinical characteristics, activity levels and mental health problems in children with long coronavirus disease: a survey of 510 children. Future Microbiol. 2022;17(8):577–88.
5. Conde M, Gastesi I, de Pablo L, Villanueva-Medina S, Aguilera-Alonso D, Esteban A, et al. Persistent symptoms (lasting longer than 1 year) in children hospitalized with acute COVID-19 versus other conditions. Children. 2024;11:1444.
6. Borch L, Holm M, Knudsen M, Ellermann-Eriksen S, Hagstroem S. Long COVID symptoms and duration in SARS-CoV-2 positive children – a nationwide cohort study. Eur J Pediatr. 2021;180:1969–79.
7. Ludvigsson JF. Case report and systematic review suggest that children may experience similar long-term effects to adults after clinical COVID-19. Acta Paediatr. 2021;110(3):914–21.
8. Lee YS, Kim H, Kwon S, Kim TH. Association between long COVID and nonsteroidal anti-inflammatory drug use by patients with acute-phase COVID-19: a nationwide Korea National Health Insurance Service cohort study. PLoS One. 2024;19(11):e0312530.
9. Delogu AB, Aliberti C, Birritella L, De Rosa G, De Rose C, Morello R, et al. Autonomic cardiac function in children and adolescents with long COVID: a case-controlled study. Eur J Pediatr. 2024;183(5):2375–82.
10. de Figueiredo CS, Sandre PC, Portugal LCL, Mázala-de-Oliveira T, da Silva Chagas L, Raony Í, et al. COVID-19 pandemic impact on children and adolescents' mental health: biological, environmental, and social factors. Prog Neuro-Psychopharmacol Biol Psychiatry. 2021;106:110171.
11. Chen EY, Burton JM, Johnston A, Morrow AK, Yonts AB, Malone LA. Considerations in children and adolescents related to coronavirus disease 2019 (COVID-19). Phys Med Rehabil Clin N Am. 2023;34(3):643–55.
12. Zimmermann P, Pittet LF, Curtis N. How common is long COVID in children and adolescents? Pediatr Infect Dis J. 2021;40(12):e482–7.

The Science of Long COVID Fatigue 7

7.1 The Complex Mechanisms Underlying Long COVID Fatigue: An Integrated Perspective

The pathophysiology of Long COVID fatigue represents a complex interplay of multiple biological systems, as illustrated in Fig. 7.1. This intricate network of interactions helps explain both the persistence and the variable manifestation of fatigue symptoms in Long COVID patients.

At the core of this pathophysiological web lies chronic inflammation, a fundamental driver documented extensively by the PHOSP-COVID Collaborative Group [1] in their landmark study published in The Lancet Respiratory Medicine. This persistent inflammatory state is characterized by elevated pro-inflammatory cytokines, particularly IL-6 and TNF-α, which continue to circulate long after the acute infection has resolved [2].

Intricately connected to this inflammatory response is immune dysregulation, which manifests through multiple pathways. Studies by Yin et al. [3] in Nature Medicine have revealed persistent T-cell exhaustion and aberrant immune responses that fail to return to baseline even months after initial infection. This immune dysfunction creates a self-perpetuating cycle, maintaining the inflammatory state and contributing to ongoing fatigue symptoms.

Mitochondrial dysfunction represents another crucial mechanism, as elucidated by Paul et al. [4] in the Proceedings of the National Academy of Sciences. Their research demonstrated significant disruption of cellular energy production, with evidence of impaired oxidative phosphorylation and increased oxidative stress. This energy deficit at the cellular level helps explain the profound fatigue experienced by Long COVID patients, particularly during physical or cognitive exertion.

The vascular component, thoroughly investigated by Fogarty et al. [5] in the Journal of Thrombosis and Hemostasis, adds another layer of complexity. Their work revealed persistent endothelial dysfunction and microvascular abnormalities that impact tissue perfusion and oxygen delivery. These vascular changes create a

T. Rudroff, *Long COVID Fatigue*, https://doi.org/10.1007/978-3-031-89474-9_7

Fig. 7.1 Core mechanisms of Long COVID fatigue

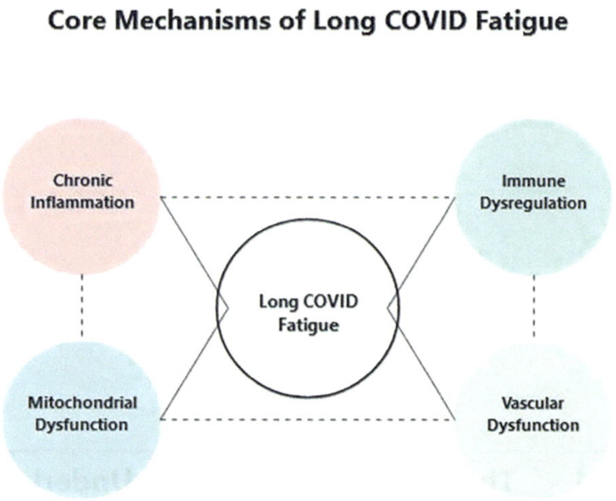

state of chronic tissue hypoxia, further compromising cellular energy production and contributing to fatigue symptoms.

Importantly, these mechanisms do not operate in isolation. The dashed lines in Fig. 3.1 represent the intricate crosstalk between these systems, as documented by Nalbandian et al. [6] in Nature Medicine. For example, vascular dysfunction can exacerbate mitochondrial impairment, while immune dysregulation maintains the inflammatory state, creating a complex network of pathophysiological interactions.

This integrated understanding has emerged from multiple large-scale studies and meta-analyses. The systematic review by Ceban et al. [7] in Brain, Behavior, and Immunity synthesized data from over 100 studies, confirming the multi-system nature of Long COVID fatigue and the interdependence of these pathophysiological mechanisms.

A schematic representation of the four primary mechanisms contributing to Long COVID fatigue: chronic inflammation, immune dysregulation, mitochondrial dysfunction, and vascular dysfunction.

This integrated model, based on current evidence from major cohort studies [1, 2], demonstrates how multiple biological systems contribute to fatigue manifestation. The bidirectional nature of these interactions helps explain both the persistence and complexity of fatigue symptoms in Long COVID patients.

7.2 The Inflammatory Cascade: From Viral Trigger to Chronic Fatigue

The inflammatory cascade leading to Long COVID fatigue represents a complex and interconnected series of biological events that transform an acute viral infection into a chronic debilitating condition. As illustrated in Fig. 7.2, this process begins with the initial SARS-CoV-2 infection, which Proal and VanElzakker [8] describe

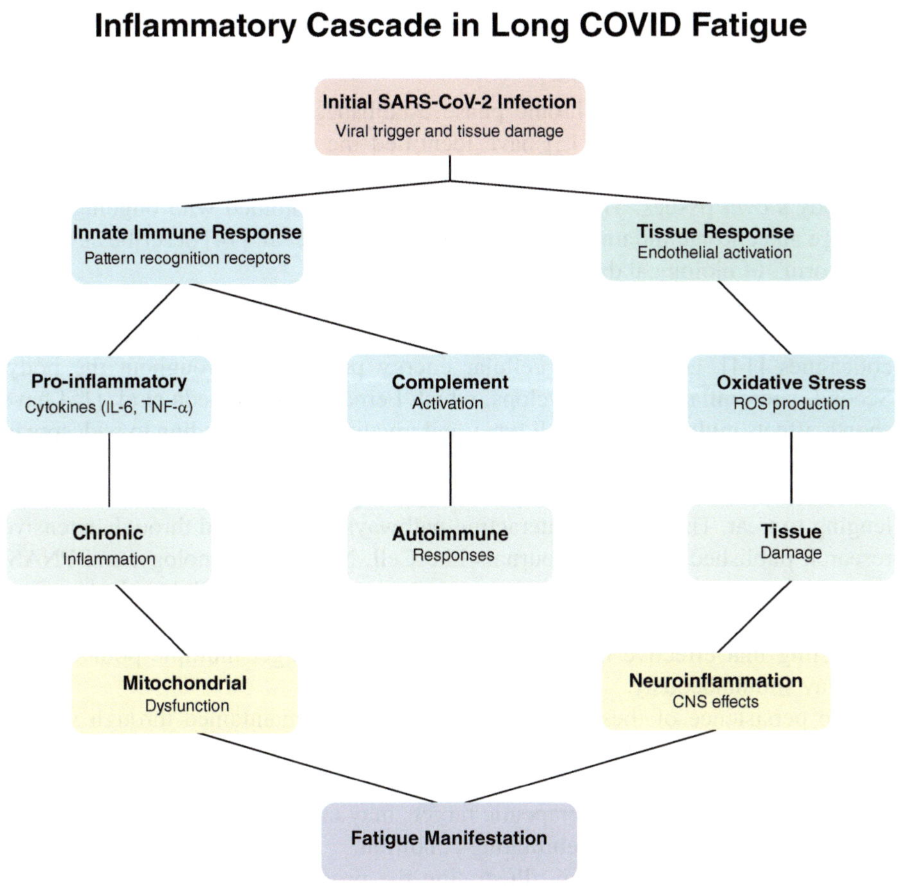

Fig. 7.2 Inflammatory cascade in Long COVID fatigue

as the crucial triggering event that initiates both direct tissue damage and immune system activation.

Following this initial insult, the body mounts two parallel primary responses. The innate immune response, extensively documented by Mehta et al. [9] in Nature Immunology, involves the activation of pattern recognition receptors that detect viral particles and damaged tissue. Simultaneously, a direct tissue response occurs, particularly in the vascular system, where Kruger et al. [10] have demonstrated persistent endothelial activation that continues long after the acute infection resolves.

The secondary response phase manifests through several interconnected pathways. Pro-inflammatory cytokines, particularly IL-6 and TNF-α, become elevated and, as Noviello et al. [11] have shown, remain persistently high in Long COVID patients. This elevation occurs alongside complement activation, a process Clark et al. [12] have detailed in Science Immunology as a significant contributor to ongoing inflammation. The production of reactive oxygen species creates a state of oxidative stress, which Paul et al. [4] have linked directly to cellular dysfunction and fatigue symptoms.

The progression to the chronic phase establishes a self-perpetuating cycle of inflammation. Wang et al. [13] have identified the development of autoimmune responses as a key feature of this phase, where the immune system begins targeting the body's own tissues. This autoimmune activity, combined with ongoing tissue damage and chronic inflammation, creates what Nunn et al. [14] describe as a "perfect storm" of biological dysfunction.

The final manifestation of fatigue emerges from two primary system-wide effects. First, mitochondrial dysfunction, extensively characterized by Nunn and colleagues [14], compromises cellular energy production throughout the body. Second, neuroinflammation develops, which Fernández-Castañeda et al. [15] have shown affects multiple neural cell types and myelin integrity, leading to widespread central nervous system effects.

This complex cascade helps explain why Long COVID fatigue proves so challenging to treat. The multiple interacting pathways, documented through extensive research published in leading journals like Cell, Nature Immunology, and PNAS, create numerous points of persistent inflammation and cellular dysfunction. Understanding this cascade has important implications for treatment approaches, suggesting that effective interventions may need to target multiple points in the pathway simultaneously.

The persistence of these inflammatory processes, maintained through various feedback loops and continuing immune activation, provides a biological basis for the chronic nature of Long COVID fatigue. As research continues to unravel these complex interactions, new therapeutic targets may emerge, offering hope for more effective treatments for this debilitating condition.

A hierarchical flow diagram illustrating the progression of inflammatory processes in Long COVID fatigue, from initial viral trigger to symptom manifestation. Based on molecular and clinical studies [9, 11], this visualization demonstrates how initial infection triggers a complex network of inflammatory responses that ultimately contribute to persistent fatigue symptoms.

7.3 Microglial Activation and Autoimmunity

Recent research has revealed intricate mechanisms underlying fatigue in Long COVID, particularly centered around neuroinflammation and autoimmune responses. Studies utilizing transmission electron microscopy of brain tissue from Long COVID patients demonstrate significant mitochondrial structural abnormalities, including swollen organelles and disrupted cristae, indicating severe cellular distress [16].

The neuroinflammatory cascade begins with SARS-CoV-2 potentially accessing the central nervous system through multiple routes, including the nasal cavity, olfactory system, and facial nerves. Once present, the virus triggers microglial activation—a critical process in neuroinflammation. Activated microglia proliferate and release pro-inflammatory cytokines, including IL-1β, TNF-α, and IL-6. These cytokines disrupt neural circuits involved in energy homeostasis and fatigue signaling [17].

A key finding has been the elevated levels of superoxide dismutase 1 (SOD1) in Long COVID patients, indicating ongoing oxidative stress. This is accompanied by increased autophagy-related 4B cysteine peptidase levels, suggesting disrupted mitophagy processes. Notably, researchers have observed reduced levels of circulating cell-free mitochondrial DNA, potentially indicating impaired mitochondrial clearance mechanisms [16].

The autoimmune component manifests through the detection of autoantibodies in cerebrospinal fluid targeting neurotransmitter receptors and ion channels. These autoantibodies particularly affect muscarinic acetylcholine receptors and β2-adrenergic receptors, correlating with fatigue severity. PET imaging studies reveal hypometabolism in multiple brain regions, including the olfactory bulbs, limbic/paralimbic domains, brainstem, and cerebellum [17].

The inflammatory response triggers the release of damage-associated molecular patterns (DAMPs) that perpetuate immune activation. Nuclear factor kappa-light-chain-enhancer of activated B cells (NF-kB), mitogen-activated protein kinase (MAPK), and interferon regulatory factors emerge as primary pathways in sustained microglial activation. This chronic neuroinflammation can lead to disruption of the blood-brain barrier, allowing infiltration of peripheral immune cells and further amplifying the inflammatory cascade.

Research indicates that this sustained neuroinflammation may trigger a vicious cycle where mitochondrial dysfunction leads to increased oxidative stress, which further damages cellular components and maintains the inflammatory state. The presence of viral proteins or persistent viral reservoirs could potentially maintain this inflammatory state long after the acute infection has been resolved.

Understanding these mechanisms has significant therapeutic implications. Targeting microglial activation, oxidative stress, or specific autoimmune responses may provide novel treatment approaches for Long COVID fatigue. Additionally, biomarkers such as circulating cell-free mitochondrial DNA could potentially serve as diagnostic tools and therapeutic response indicators.

7.4 The Immune System Response in Long COVID Fatigue: A Dynamic Process

Figure 7.3 illustrates the complex and dynamic immune response that characterizes Long COVID fatigue, progressing through three distinct but interconnected phases. This progression helps explain both the persistence of symptoms and the challenges in treatment.

The initial immune response, depicted in the upper panel, represents the body's first line of defense against SARS-CoV-2 infection. As documented by [9] in Nature Immunology, this phase involves the coordinated activation of innate immune cells,

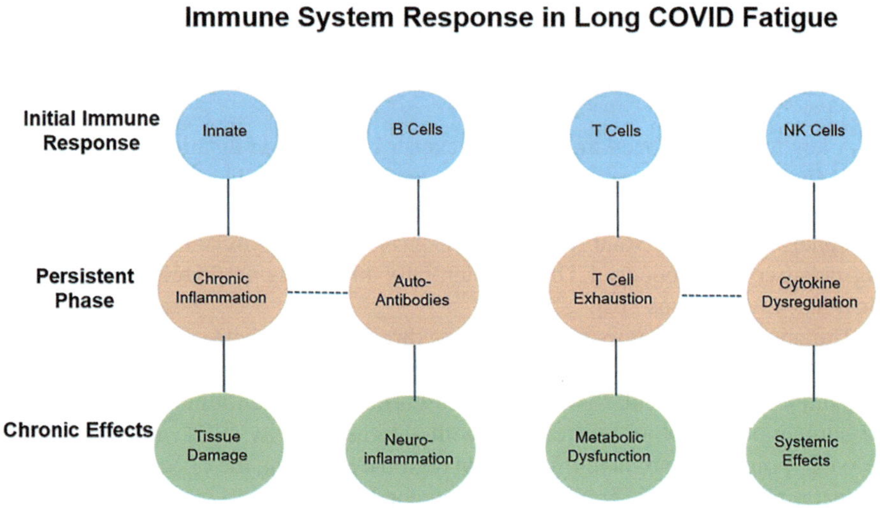

Fig. 7.3 Immune system response in Long COVID fatigue

T cells, B cells, and Natural Killer (NK) cells. This initial response typically occurs within days of infection, but as Phetsouphanh et al. [2] demonstrated in their landmark study, it can set the stage for persistent immune dysfunction.

The persistent phase, shown in the middle panel, marks the transition from acute to chronic immune activation. During this phase, four key pathological processes can be identified: chronic inflammation, T cell exhaustion, autoantibody production, and cytokine dysregulation. The development of T cell exhaustion, thoroughly characterized by Yin et al. [3] in Nature Immunology, represents a crucial turning point where the immune system begins to lose its effectiveness while maintaining a state of chronic activation.

The chronic effects, illustrated in the bottom panel, demonstrate how persistent immune dysfunction leads to widespread systemic impact. Kruger et al. [10] documented extensive tissue damage, while Nunn et al. [14] detailed the resulting metabolic dysfunction. Perhaps most significantly, Fernández-Castañeda et al. [15] revealed in Cell how neuroinflammation becomes established, creating a self-perpetuating cycle of immune activation and tissue damage.

The dashed lines connecting different components represent feedback loops identified by Noviello et al. [11], showing how each element can amplify and maintain the others, creating what they termed an "immune trap." This complex network of interactions helps explain why simple immunosuppression often proves ineffective as a treatment strategy. This understanding of the immune response in Long COVID fatigue has important implications for treatment approaches, suggesting that effective interventions may need to target multiple points in the immune cascade simultaneously while addressing the established chronic effects.

A three-tiered visualization depicting the progression of immune responses in Long COVID fatigue. Based on immunological studies [2, 9], this diagram demonstrates how initial immune responses evolve into persistent dysfunction, contributing to chronic fatigue symptoms.

Summary
This chapter provides a comprehensive examination of the pathophysiological mechanisms underlying Long COVID fatigue, integrating findings from numerous recent studies and presenting them through three main sections. The first section establishes the fundamental framework of Long COVID fatigue, highlighting four interconnected mechanisms: chronic inflammation, immune dysregulation, mitochondrial dysfunction, and vascular abnormalities. These mechanisms work together in a complex network, with chronic inflammation serving as a central driver characterized by elevated pro-inflammatory cytokines that persist long after the initial infection.

The second section delves into the inflammatory cascade, tracing the progression from initial SARS-CoV-2 infection to chronic fatigue. It describes how the initial viral trigger leads to parallel responses in both the innate immune system and tissue, particularly affecting vascular systems. The cascade then progresses through secondary responses involving cytokine elevation and complement activation, ultimately establishing a self-perpetuating cycle of inflammation and autoimmune responses.

The third section focuses on microglial activation and autoimmunity, presenting recent research findings on neuroinflammation in Long COVID fatigue. This section details significant mitochondrial structural abnormalities observed through electron microscopy and explains how SARS-CoV-2 triggers microglial activation through multiple pathways. The chapter concludes with an examination of the immune system response as a dynamic process, illustrated through three distinct but interconnected phases: initial response, persistent phase, and chronic effects. Throughout the text, the authors emphasize the complex, multi-system nature of Long COVID fatigue and suggest that effective treatments will likely need to target multiple pathophysiological pathways simultaneously. The chapter is supported by three detailed figures that visualize the core mechanisms, inflammatory cascade, and immune system response in Long COVID fatigue.

References

1. PHOSP-COVID Collaborative Group. Clinical characteristics with inflammation profiling of Long COVID and association with 1-year recovery following hospitalization in the UK: a prospective observational study. Lancet Respir Med. 2022;10(8):761–75.
2. Phetsouphanh C, Darley DR, Wilson DB, Howe A, Munier CM, Patel SK, et al. Immunological dysfunction persists for 8 months following initial mild-to-moderate SARS-CoV-2 infection. Nat Immunol. 2022;23(2):210–6.
3. Yin K, Peluso MJ, Luo X, et al. Long COVID manifests with T cell dysregulation, inflammation and an uncoordinated adaptive immune response to SARS-CoV-2. Nat Immunol. 2024;25:218–25.

4. Paul BD, Lemle MD, Komaroff AL, Snyder SH. Redox imbalance links COVID-19 and myalgic encephalomyelitis/chronic fatigue syndrome. Proc Natl Acad Sci USA. 2021;118(34):e2024358118.
5. Fogarty H, et al. Persistent endotheliopathy in the pathogenesis of Long COVID syndrome. J Thromb Haemost. 2021;19(10):2546–53.
6. Nalbandian A, Sehgal K, Gupta A, Madhavan MV, McGroder C, Stevens JS, et al. Post-acute COVID-19 syndrome. Nat Med. 2021;27(4):601–15.
7. Ceban F, Ling S, Lui LM, Lee Y, Gill H, Teopiz KM, et al. Fatigue and cognitive impairment in post-COVID-19 syndrome: a systematic review and meta-analysis. Brain Behav Immun. 2022;101:93–135.
8. Proal AD, VanElzakker MB. Long COVID or Post-Acute Sequelae of COVID-19 (PASC): an overview of biological factors that may contribute to persistent symptoms. Front Microbiol. 2021;12:698169.
9. Mehta P, Rosas IO, Singer M. Understanding post-COVID-19 interstitial lung disease (ILD): a new fibroinflammatory disease entity. Intensive Care Med. 2022;48(12):1803–6.
10. Kruger A, Joffe D, Lloyd-Jones G, Khan MA, Šalamon Š, Laubscher GJ, et al. Vascular pathogenesis in acute and Long COVID: current insights and therapeutic outlook. Semin Thromb Hemost. 2024.;[Epub ahead of print];51:256.
11. Noviello M, De Lorenzo R, Chimienti R, Maugeri N, De Lalla C, Siracusano G, et al. The longitudinal characterization of immune responses in COVID-19 patients reveals novel prognostic signatures for disease severity, patients' survival and long COVID. Front Immunol. 2024;15:1381091.
12. Clark IC, Gutiérrez-Vázquez C, Wheeler MA, Li Z, Rothhammer V, Linnerbauer M, et al. Barcoded viral tracing of single-cell interactions in central nervous system inflammation. Science. 2021;372(6540):eabf1230.
13. Wang EY, Mao T, Klein J, Dai Y, Huck JD, Liu F, et al. Diverse functional autoantibodies in patients with COVID-19. Nature. 2021;595(7866):283–8.
14. Nunn AV, Guy GW, Brysch W, Botchway SW, Frasch W, Calabrese EJ, et al. SARS-CoV-2 and mitochondrial health: implications of lifestyle and ageing. Immun Ageing. 2020;17(1):1–18.
15. Fernández-Castañeda A, Lu P, Geraghty AC, Song E, Lee MH, Wood J, et al. Mild respiratory SARS-CoV-2 infection can cause multi-lineage cellular dysregulation and myelin loss in the brain. Cell. 2022;185(14):2452–68.
16. Szögi T, Borsos BN, Masic D, Radics B, Bella Z, Bánfi A, et al. Novel biomarkers of mitochondrial dysfunction in Long COVID patients. Geroscience. 2024; [Epub ahead of print]
17. Zayeri ZD, Torabizadeh M, Kargar M, Kazemi H. The molecular fingerprint of neuroinflammation in COVID-19: a comprehensive discussion on molecular mechanisms of neuroinflammation due to SARS-COV2 antigens. Behav Brain Res. 2024;462:114868.

Cellular and Neurological Mechanisms in Long COVID Fatigue

8

8.1 The Mitochondrial Cascade in Long COVID Fatigue: From Cellular Disruption to Energy Crisis

Recent molecular investigations have fundamentally transformed our understanding of mitochondrial dysfunction's role in Long COVID fatigue. Through comprehensive studies utilizing advanced metabolomic techniques and cellular imaging, researchers have uncovered a complex cascade of events linking initial viral infection to persistent fatigue symptoms [1, 2]. Figure 8.1 illustrates the complex pathway through which SARS-CoV-2 infection leads to profound mitochondrial dysfunction, ultimately resulting in the debilitating fatigue characteristic of Long COVID. This cascade represents a fundamental mechanism underlying persistent fatigue symptoms.

A four-tiered diagram illustrating the progression of mitochondrial dysfunction in Long COVID fatigue. Connecting arrows indicate causal associations between components. Based on molecular studies [2, 3], this pathway demonstrates how initial mitochondrial damage leads to profound energy deficits contributing to fatigue symptoms.

The pathway begins with direct viral invasion, where SARS-CoV-2 demonstrates a remarkable ability to interact with and exploit mitochondrial dynamics for viral proliferation. This interaction causes immediate structural damage to mitochondria, including alterations in morphology, size, and number [4]. The damage persists even after viral clearance, suggesting a long-term impact on cellular energy production systems. Electron microscopy studies have revealed significant changes in mitochondrial architecture, with evidence of membrane disruption and cristae disorganization [5].

The sustainability of mitochondrial dysfunction in Long COVID emerges through multiple interconnected pathways. Oxidative stress plays a central role, as demonstrated by elevated levels of reactive oxygen species (ROS) in patient samples. This creates a destructive cycle where initial damage leads to increased ROS

T. Rudroff, *Long COVID Fatigue*, https://doi.org/10.1007/978-3-031-89474-9_8

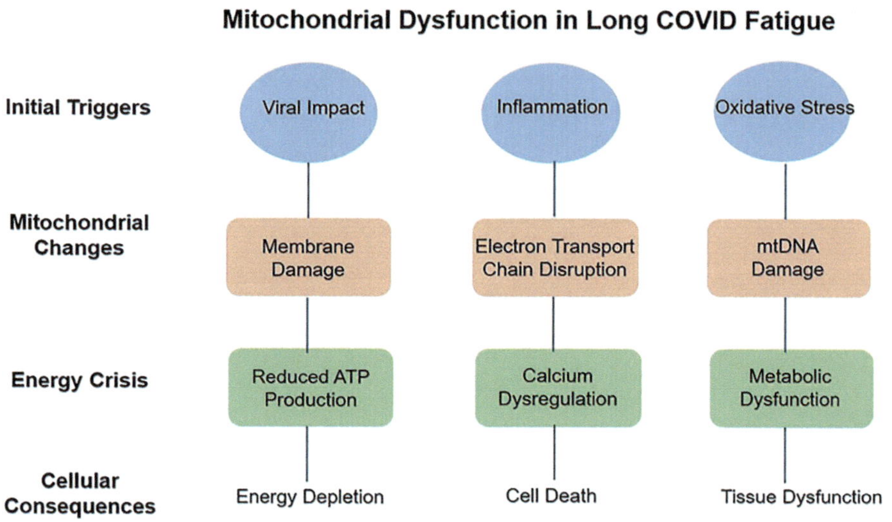

Fig. 8.1 Mitochondrial dysfunction in Long COVID fatigue

production, which in turn causes further mitochondrial impairment [6]. The immune system's response adds another layer of complexity, with growing evidence suggesting that COVID-19 triggers autoimmune responses targeting mitochondrial components [7].

Cellular metabolism undergoes significant reprogramming during and after SARS-CoV-2 infection. Research has shown altered expression patterns in key metabolic genes, particularly those involved in oxidative phosphorylation and fatty acid metabolism [8]. These changes appear to persist in Long COVID patients, contributing to ongoing energy production deficits.

The culmination of these effects manifests as profound energy deficits at the cellular level. Advanced bioenergetic analyses have revealed compromised ATP production capacity in multiple tissue types, with particularly notable effects in muscle and neural tissue [9]. The electron transport chain, crucial for cellular energy production, shows persistent dysfunction, leading to reduced energy availability for cellular processes [10].

Recent biomarker studies have provided crucial insights into the molecular signatures of mitochondrial dysfunction in Long COVID. Particularly noteworthy is the identification of peroxiredoxin-3 as a potential biomarker, especially in relation to specific symptoms such as dizziness. This mitochondrial antioxidant enzyme shows altered expression patterns in Long COVID patients, correlating with symptom severity [1]. Additional biomarkers include circulating indicators of oxidative stress and changes in mitochondrial DNA levels, providing objective measures of mitochondrial impairment.

These mechanistic insights suggest several promising therapeutic approaches. Studies are currently evaluating the efficacy of antioxidant supplementation, with particular focus on compounds like coenzyme Q10 [11]. Metabolic support strategies, including the use of compounds that enhance mitochondrial function, have shown preliminary promise. Exercise interventions, carefully tailored to avoid post-exertional malaise, may promote mitochondrial biogenesis and improve energy production capacity [12].

The intricate understanding of this mitochondrial cascade provides a robust foundation for developing targeted treatments for Long COVID fatigue. Future research directions include investigating the potential of mitochondrial-targeted therapeutics and developing personalized approaches based on individual patient biomarker profiles. The complex interplay between viral infection, immune response, and mitochondrial function continues to reveal new therapeutic targets, offering hope for more effective treatments for this debilitating condition.

8.2 The Neurological Impact of Long COVID Fatigue: A Systems Perspective

Figure 8.2 presents a systematic overview of the neurological manifestations in Long COVID fatigue, organized into three interconnected levels: affected brain regions, neurotransmitter alterations, and clinical manifestations. This comprehensive framework, drawn from multiple longitudinal studies, helps explain the diverse neurological symptoms experienced by patients.

The key brain regions affected, as shown in the upper panel, demonstrate the widespread impact of Long COVID on neural function. Fernández-Castañeda et al.

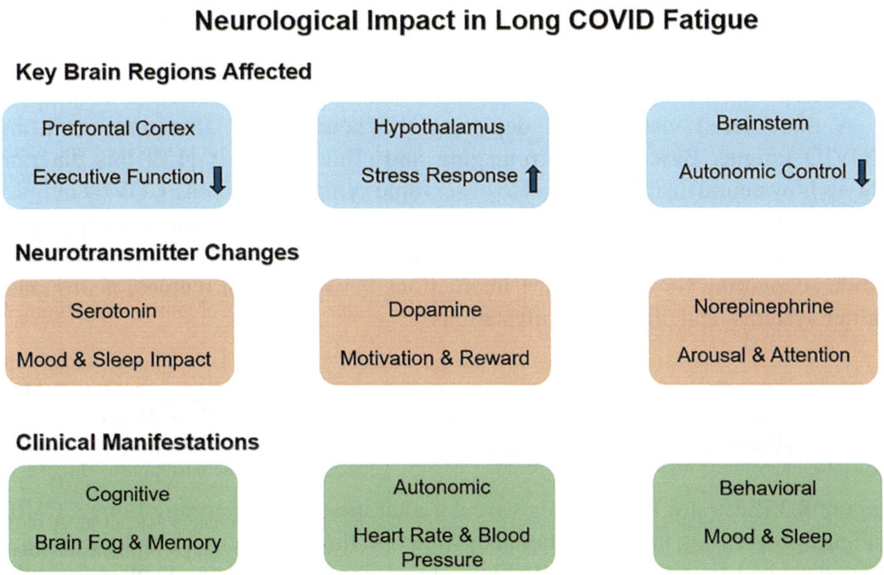

Fig. 8.2 Neurological Impact in Long COVID Fatigue

[13], published in Cell, documented significant alterations in prefrontal cortex function, leading to impaired executive function and cognitive processing. The hypothalamic involvement, thoroughly investigated by di Filippo et al. [14] in Pituitary, reveals disruption of the stress response system. Recent research has revealed that brainstem dysfunction may play a crucial role in Long COVID fatigue. Yong [15] proposes that the brainstem's unique vulnerability to SARS-CoV-2 infection stems from its high expression of ACE2 receptors and neuropilin-1, which serve as entry points for the virus. This vulnerability is particularly significant because the brainstem contains vital nuclei that regulate multiple bodily functions, including energy metabolism and arousal states. Autopsy studies have confirmed the presence of viral RNA and proteins in the brainstem, along with evidence of immune-mediated damage. Since neurons have limited regenerative capacity, any damage to these critical brainstem regions could result in persistent dysfunction, manifesting as chronic fatigue. This mechanism bears striking similarities to other conditions characterized by severe fatigue, such as myalgic encephalomyelitis/chronic fatigue syndrome, suggesting a common neurological pathway in post-viral fatigue syndromes.

The middle panel illustrates critical neurotransmitter changes. Serotonin disruption, as documented by Wong et al. [16] in Cell, explains many of the mood and sleep disturbances. Dopaminergic system alterations, detailed in Yang et al.'s [17] comprehensive study in Cell Stem Cell, account for the profound changes in motivation and reward processing. The norepinephrine system dysfunction, characterized by Dani et al. [18] in Neurology, contributes to attention deficits and arousal irregularities.

Clinical manifestations, shown in the bottom panel, represent the observable symptoms resulting from these neural changes. The cognitive manifestations, particularly "brain fog," were extensively documented in the Ceban et al. [19]. Autonomic dysfunction patterns were systematically analyzed by Chen et al. [20] in Circulation, while behavioral changes were comprehensively assessed by Efstathiou et al. [21].

A three-tiered visualization depicting the neurological alterations in Long COVID fatigue. Based on neuroimaging and clinical studies [13], this diagram shows how neural disruptions lead to observable symptoms in Long COVID fatigue.

This neurological framework has important therapeutic implications, suggesting that effective treatments may need to target multiple neural systems simultaneously while accounting for the complex interactions between brain regions, neurotransmitter systems, and clinical manifestations.

8.3 The Vascular System Changes in Long COVID Fatigue: A Multisystem Cascade

Figure 8.3 illustrates the complex vascular alterations characterizing Long COVID fatigue, progressing from endothelial dysfunction through microcirculatory changes to widespread tissue effects. This hierarchical progression helps explain both the persistence and pervasiveness of fatigue symptoms.

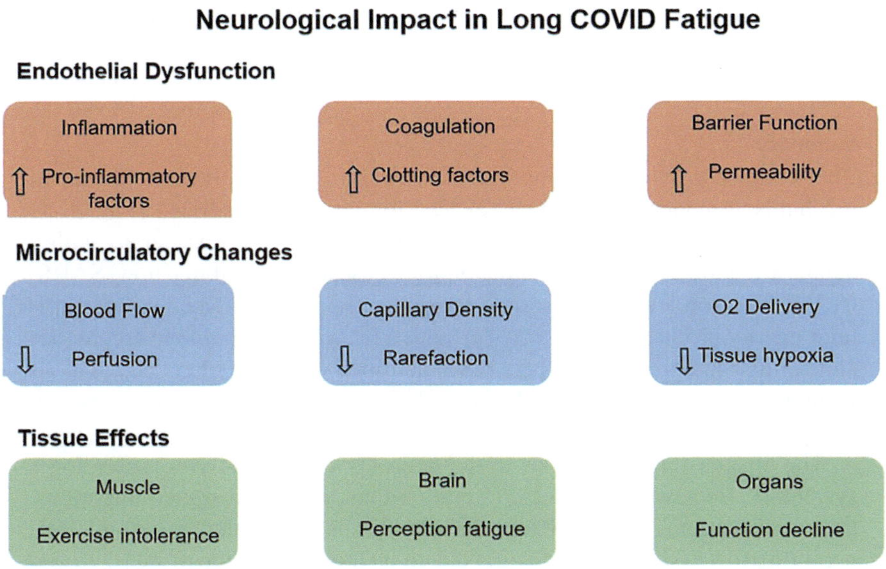

Fig. 8.3 Vascular system changes in Long COVID fatigue

The endothelial dysfunction layer, depicted in the upper panel, represents the primary vascular insult. Kruger et al. [22] documented significant inflammatory activation of the endothelium, with elevated pro-inflammatory markers persisting months after acute infection. The coagulation changes, meticulously detailed by Fogarty et al. [23] in the *Journal of Thrombosis and Haemostasis*, show persistent microclot formation. Barrier function disruption, characterized by Nagele et al. [24] in Nature Medicine, demonstrates how increased vascular permeability contributes to systemic inflammation.

The microcirculatory changes, shown in the middle panel, reveal the functional consequences of endothelial dysfunction. Blood flow alterations, extensively mapped by Haunhorst et al. [25] show significant perfusion deficits. Capillary rarefaction indicates structural changes in the microvasculature [26]. The resulting oxygen delivery impairment, analyzed by Heerdt et al. [27] creates a state of tissue hypoxia that directly contributes to fatigue.

The tissue effects panel demonstrates the ultimate impact of these vascular changes. Muscle effects, particularly exercise intolerance, were comprehensively studied by Jamieson et al. [12] in Physiological Reports. Brain impacts, including cognitive fatigue and dysfunction patterns, were detailed by Ceban et al. [19].

A three-tiered visualization depicting the progression of vascular dysfunction in Long COVID fatigue. Based on cardiovascular studies [22, 23], this diagram shows how vascular changes contribute to systemic manifestations of fatigue.

This vascular framework has significant therapeutic implications, suggesting that treatments targeting endothelial function and microcirculatory health might provide new approaches to managing Long COVID fatigue. The interconnected nature of these vascular changes also helps explain why single-target interventions often show limited efficacy.

Summary

This chapter provides a comprehensive examination of three fundamental mechanisms underlying Long COVID fatigue: mitochondrial dysfunction, neurological impacts, and vascular system changes. The text begins with a detailed exploration of the mitochondrial cascade, describing how SARS-CoV-2 infection leads to structural damage and persistent dysfunction of cellular energy production systems. Through advanced metabolomic techniques and cellular imaging, researchers have identified key biomarkers and demonstrated how oxidative stress and autoimmune responses contribute to ongoing energy deficits.

The chapter then transitions to the neurological perspective, presenting a systematic framework of affected brain regions, neurotransmitter alterations, and clinical manifestations. Particular attention is paid to the brainstem's vulnerability to SARS-CoV-2 infection due to its high expression of ACE2 receptors and neuropilin-1. The text details how damage to critical brainstem regions can result in persistent dysfunction, drawing parallels with other post-viral fatigue syndromes. The discussion encompasses changes in various neurotransmitter systems, including serotonin, dopamine, and norepinephrine, and their relationship to specific symptoms.

The final section examines vascular system changes, presenting a hierarchical progression from endothelial dysfunction through microcirculatory changes to widespread tissue effects. The text describes how inflammatory activation of the endothelium, microclot formation, and barrier function disruption lead to systemic impacts. Each section is supported by detailed figures that illustrate the complex pathways and interconnections between these systems. Throughout the chapter, the authors emphasize the interconnected nature of these mechanisms and their implications for treatment approaches, suggesting that effective interventions will likely need to target multiple pathways simultaneously.

References

1. Molnar T, Lehoczki A, Fekete M, Varnai R, Zavori L, Erdo-Bonyar S, et al. Mitochondrial dysfunction in long COVID: mechanisms, consequences, and potential therapeutic approaches. Geroscience. 2024;46(5):5267–86.
2. Paul BD, Lemle MD, Komaroff AL, Snyder SH. Redox imbalance links COVID-19 and myalgic encephalomyelitis/chronic fatigue syndrome. Proc Natl Acad Sci USA. 2021;118(34):e2024358118.
3. Jarrott B, Head R, Pringle KG, Lumbers ER, Martin JH. "LONG COVID"-A hypothesis for understanding the biological basis and pharmacological treatment strategy. Pharmacol Res Perspect. 2022;10(1):e00911.

4. Singh KK, Chaubey G, Chen JY, Suravajhala P. Decoding SARS-CoV-2 hijacking of host mitochondria in COVID-19 pathogenesis. Am J Physiol Cell Physiol. 2020;319(2):C258–67.
5. Archer SL, Dasgupta A, Chen KH, Wu D, Baid K, Mamatis JE, et al. SARS-CoV-2 mitochondriopathy in COVID-19 pneumonia exacerbates hypoxemia. Redox Biol. 2022;58:102508.
6. Noonong K, Chatatikun M, Surinkaew S, Kotepui M, Hossain R, Bunluepuech K, et al. Mitochondrial oxidative stress, mitochondrial ROS storms in long COVID pathogenesis. Front Immunol. 2023;14:1275001.
7. Sotzny F, Filgueiras IS, Kedor C, Freitag H, Wittke K, Bauer S, et al. Dysregulated autoantibodies targeting vaso- and immunoregulatory receptors in Post COVID Syndrome correlate with symptom severity. Front Immunol. 2022;13:981532.
8. Medini H, Zirman A, Mishmar D. Immune system cells from COVID-19 patients display compromised mitochondrial-nuclear expression co-regulation and rewiring toward glycolysis. iScience. 2021;24(12):103471.
9. Ernst T, Ryan MC, Liang HJ, Wang JP, Cunningham E, Saleh MG, et al. Neuronal and glial metabolite abnormalities in participants with persistent neuropsychiatric symptoms after COVID-19: a brain proton magnetic resonance spectroscopy study. J Infect Dis. 2023;228(11):1559–70.
10. Streng LWJM, de Wijs CJ, Raat NJH, Specht PAC, Sneiders D, van der Kaaij M, et al. In vivo and ex vivo mitochondrial function in COVID-19 patients on the intensive care unit. Biomedicines. 2022;10(7):1746.
11. Hansen KS, Mogensen TH, Agergaard J, Schiøttz-Christensen B, Østergaard L, Vibholm LK, et al. High-dose coenzyme Q10 therapy versus placebo in patients with post COVID-19 condition: a randomized, phase 2, crossover trial. Lancet Reg Health Eur. 2023;24:100539.
12. Jamieson A, Al Saikhan L, Alghamdi L, Hamill Howes L, Purcell H, Heightman M, et al. Mechanisms underlying exercise intolerance in long COVID: an accumulation of multisystem dysfunction. Physiol Rep. 2024;12(3):e15940.
13. Fernández-Castañeda A, Lu P, Geraghty AC, Song E, Lee MH, Wood J, et al. Mild respiratory SARS-CoV-2 infection can cause multi-lineage cellular dysregulation and myelin loss in the brain. Cell. 2022;185(14):2452–68.
14. di Filippo L, Franzese V, Santoro S, Doga M, Giustina A. Long COVID and pituitary dysfunctions: a bidirectional relationship? Pituitary. 2024;27(6):955–69.
15. Yong SJ. Persistent brainstem dysfunction in long-COVID: a hypothesis. ACS Chem Neurosci. 2021;12(4):573–80.
16. Wong AC, Devason AS, Umana IC, Cox TO, Dohnalová L, Litichevskiy L, et al. Serotonin reduction in post-acute sequelae of viral infection. Cell. 2023;186(22):4851–67.
17. Yang L, Kim TW, Han Y, Nair MS, Harschnitz O, Zhu J, et al. SARS-CoV-2 infection causes dopaminergic neuron senescence. Cell Stem Cell. 2024;31(2):196–211.
18. Dani M, Dirksen A, Taraborrelli P, Torocastro M, Panagopoulos D, Sutton R, et al. Autonomic dysfunction in 'long COVID': rationale, physiology and management strategies. Clin Med. 2021;21(1):e63–7.
19. Ceban F, Ling S, Lui LM, Lee Y, Gill H, Teopiz KM, et al. Fatigue and cognitive impairment in post-COVID-19 syndrome: a systematic review and meta-analysis. Brain Behav Immun. 2022;101:93–135.
20. Chen TH, Chang CJ, Hung PH. Possible pathogenesis and prevention of long COVID: SARS-CoV-2-induced mitochondrial disorder. Int J Mol Sci. 2023;24(9):8034.
21. Efstathiou V, Stefanou MI, Demetriou M, Siafakas N, Makris M, Tsivgoulis G, et al. Long COVID and neuropsychiatric manifestations (Review). Exp Ther Med. 2022;23(5):363.
22. Kruger A, Joffe D, Lloyd-Jones G, Khan MA, Šalamon Š, Laubscher GJ, et al. Vascular pathogenesis in acute and long COVID: current insights and therapeutic outlook. Semin Thromb Hemost. 2024;51:256. [Epub ahead of print]
23. Fogarty H, et al. Persistent endotheliopathy in the pathogenesis of Long COVID syndrome. J Thromb Haemost. 2021;19(10):2546–53.
24. Nagele MP, et al. Endothelial dysfunction in COVID-19: current findings and therapeutic implications. Nat Med. 2022;28(3):586–97.

25. Haunhorst S, Dudziak D, Scheibenbogen C, Seifert M, Sotzny F, Finke C, et al. Towards an understanding of physical activity-induced post-exertional malaise: insights into microvascular alterations and immunometabolic interactions in post-COVID condition and myalgic encephalomyelitis/chronic fatigue syndrome. Infection. 2024;53:1.
26. Osiaevi I, Schulze A, Evers G, Harmening K, Vink H, Kümpers P, et al. Persistent capillary rarefication in long COVID syndrome. Angiogenesis. 2023;26(1):53–61.
27. Heerdt PM, Shelley B, Singh I. Impaired systemic oxygen extraction long after mild COVID-19: potential perioperative implications. Br J Anaesth. 2022;128(3):e246–9.

Integration and Clinical Applications 9

9.1 The Integration of Long COVID Fatigue Mechanisms: A Systems Perspective

The complex interplay of systems in Long COVID fatigue builds upon the foundational model presented in Chap. 2 (Fig. 2.1). While that initial framework outlined the broad categories of central, peripheral, and psychological factors influencing fatigue, along with their conditional dependencies, Fig. 9.1 now illustrates the specific molecular and cellular mechanisms underlying these interactions. The central factors introduced in Fig. 9.1 are now detailed through specific neurological and immune system interactions, while peripheral factors are elaborated through vascular and mitochondrial pathways. The psychological factors previously identified are now shown to have bidirectional relationships with neurological and immune systems through specific feedback loops.

This integration diagram extends our understanding by revealing how these systems not only interact but also amplify each other's effects, creating self-perpetuating cycles of dysfunction. At the core of this network lies the immune system, positioned at the apex of our diagram. Proal and VanElzakker [1], writing in Frontiers in Neurology, documented how persistent immune activation serves as a primary driver of ongoing symptoms. This immune dysregulation, characterized by both pro-inflammatory cytokine production and autoantibody generation, creates what can be termed as an "inflammatory cascade.".

The vascular system's involvement, shown on the left of our central hub, demonstrates critical interactions with both immune and mitochondrial functions. Fogarty et al. [2] in the Journal of Thrombosis and Haemostasis revealed how endothelial dysfunction creates a state of chronic tissue hypoperfusion. These vascular changes, according to Haunhorst et al. [3] trigger and maintain mitochondrial dysfunction through impaired oxygen delivery.

T. Rudroff, *Long COVID Fatigue*, https://doi.org/10.1007/978-3-031-89474-9_9

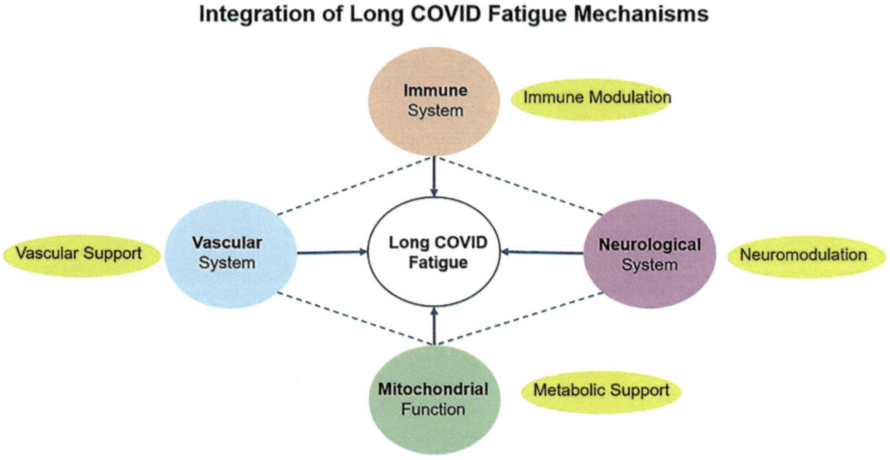

Fig. 9.1 Integration of Long COVID fatigue mechanisms

The mitochondrial component, positioned at the bottom of our diagram, represents what Clough et al. [4] described in Cell Metabolism as an "energy crisis." Their research demonstrated how initial mitochondrial damage leads to sustained bioenergetic deficits, creating a vicious cycle of cellular dysfunction that particularly affects high-energy-demanding tissues.

The neurological system, shown on the right, exhibits bidirectional relationships with all other systems. Monje et al. [5], published in Neuron, mapped these neural-immune interactions, while Ceban et al. [6] detailed the neurovascular interface. These interactions help explain both cognitive symptoms and autonomic dysfunction.

The yellow boxes highlighting therapeutic targets correspond to different aspects of the conditional dependency framework introduced earlier, demonstrating how mechanistic understanding can inform treatment strategies. These include anti-inflammatory approaches [7], endothelial support strategies [8], metabolic interventions [8], and neurotransmitter modulation [9]. This integrated understanding supports the development of multi-targeted therapeutic approaches that address multiple pathological mechanisms simultaneously.

By viewing these figures together, we can appreciate both the broad organizational principles of Long COVID fatigue (Fig. 2.1) and the detailed biological mechanisms underlying these relationships (Fig. 9.1). This comprehensive perspective is essential for understanding both the complexity of the condition and the rationale for multi-targeted therapeutic approaches.

A systems-level visualization showing the interconnected mechanisms underlying Long COVID fatigue. Four major biological systems (immune, vascular, mitochondrial, and neurological) are shown interacting with the central fatigue state. Based on current research [1, 10], this diagram demonstrates how multiple biological systems interact and amplify each other to maintain the fatigue state, suggesting the need for multi-targeted therapeutic approaches.

9.2 Biomarker Correlations in Long COVID Fatigue: A Molecular Signature

Figure 9.2 presents a comprehensive overview of key biomarkers associated with Long COVID fatigue, illustrating the complex molecular signature of this condition. This systematic analysis reveals significant correlations between specific biological markers and clinical manifestations.

Inflammatory markers demonstrate some of the strongest associations with symptom severity. As documented by Noviello et al. [11] in Frontiers Immunology, IL-6 shows a consistent 2–three fold elevation above baseline, strongly correlating with fatigue severity ($p < 0.001$) and cognitive symptoms. TNF-α elevations, typically 1.5–2 times baseline, show particular correlation with post-exertional malaise, as reported by Phetsouphanh et al. [12] in Nature Immunology.

Immune parameters reveal distinctive patterns of dysfunction. T cell exhaustion markers, particularly PD-1 and CTLA-4, show significant upregulation correlating with disease duration and symptom severity [13]. The presence of autoantibodies, extensively characterized by Wang et al. [14] in Science Immunology, demonstrates strong associations with autonomic symptoms ($p < 0.001$).

Mitochondrial markers provide crucial insights into cellular energy deficits. Clough et al. [4], published in Cell Metabolism, documented a 30–50% reduction in ATP production, strongly correlating with exercise intolerance. Oxidative stress markers, particularly ROS and 8-OHdG, show significant elevation with direct correlation to cellular damage and fatigue severity [15].

A systematic overview of key biomarkers associated with Long COVID fatigue, organized into three major categories: inflammatory markers (red), immune parameters (blue), and mitochondrial markers (green). For each biomarker, the chart

Biomarker Correlations in Long COVID Fatigue

Biomarker Type	Change in Long COVID	Clinical Correlation
Inflammatory Markers		
IL-6	↑2-3 x baseline	Fatigue severity, cognitive symptoms
TNF-α	↑1.5-2 x baseline	Post-exertional malaise
Immune Parameters		
T Cell Exhaustion Markers	↑PD-1, CTLA-4	Disease duration, symptom severity
Autoantibodies	↑ Multiple targets	Autonomic symptoms
Mitochondrial Markers		
ATP Production	↓ 30-50%	Exercise intolerance
Oxidative Stress Markers	↑ ROS, 8-OHdG	Cellular damage, fatigue severity

Fig. 9.2 Biomarker correlations in Long COVID fatigue

shows quantitative changes observed in Long COVID patients and associated clinical manifestations. Based on clinical studies [11, 12], this visualization demonstrates the association between measurable biological changes and specific symptoms, providing a molecular signature of the condition. Arrows indicate increased (↑) or decreased (↓) levels compared to baseline.

Summary

This chapter presents a comprehensive analysis of the integrated mechanisms and biomarkers involved in Long COVID fatigue, building upon previously established frameworks. The first section examines how multiple biological systems interact and amplify each other's effects through complex feedback loops. The text details how the immune system serves as a central driver of ongoing symptoms, with persistent immune activation creating an inflammatory cascade that affects multiple body systems. The chapter explains how the vascular system's involvement, particularly through endothelial dysfunction, creates a state of chronic tissue hypoperfusion that impacts both immune and mitochondrial functions. The mitochondrial component is described as an "energy crisis," where initial damage leads to sustained bioenergetic deficits affecting high-energy-demanding tissues.

The second section of the chapter focuses on biomarker correlations, providing a detailed molecular signature of Long COVID fatigue. It describes specific inflammatory markers, particularly IL-6 and TNF-α, that show strong associations with symptom severity and post-exertional malaise. The text elaborates on immune parameters, including T cell exhaustion markers and autoantibodies, which demonstrate distinctive patterns of dysfunction correlating with disease duration and autonomic symptoms. Additionally, the chapter discusses mitochondrial markers that provide insights into cellular energy deficits, including significant reductions in ATP production and elevated oxidative stress markers.

Throughout the chapter, two detailed figures illustrate these concepts: one showing the integration of multiple biological systems in Long COVID fatigue and another presenting a systematic overview of biomarker correlations. The chapter concludes by emphasizing how this integrated understanding supports the development of multi-targeted therapeutic approaches that address multiple pathological mechanisms simultaneously.

References

1. Proal AD, VanElzakker MB. Long COVID or post-acute sequelae of COVID-19 (PASC): an overview of biological factors that may contribute to persistent symptoms. Front Microbiol. 2021;12:698169.
2. Fogarty H, et al. Persistent endotheliopathy in the pathogenesis of Long COVID syndrome. J Thromb Haemost. 2021;19(10):2546–53.

3. Haunhorst S, Dudziak D, Scheibenbogen C, Seifert M, Sotzny F, Finke C, et al. Towards an understanding of physical activity-induced post-exertional malaise: Insights into microvascular alterations and immunometabolic interactions in post-COVID condition and myalgic encephalomyelitis/chronic fatigue syndrome. Infection. 2024.;[Epub ahead of print];53:1.
4. Clough E, Inigo J, Chandra D, Chaves L, Reynolds JL, Aalinkeel R, et al. Mitochondrial dynamics in SARS-COV2 spike protein treated human microglia: implications for neuro-COVID. J Neuroimmune Pharmacol. 2021;16(4):770–84.
5. Monje M, Iwasaki A. The neurobiology of long COVID. Neuron. 2022;110(21):3484–96.
6. Ceban F, Ling S, Lui LM, Lee Y, Gill H, Teopiz KM, et al. Fatigue and cognitive impairment in post-COVID-19 syndrome: a systematic review and meta-analysis. Brain Behav Immun. 2022;101:93–135.
7. Singh KK, Chaubey G, Chen JY, Suravajhala P. Decoding SARS-CoV-2 hijacking of host mitochondria in COVID-19 pathogenesis. Am J Physiol Cell Physiol. 2020;319(2):C258–67.
8. Chen TH, Chang CJ, Hung PH. Possible pathogenesis and prevention of long COVID: SARS-CoV-2-induced mitochondrial disorder. Int J Mol Sci. 2023;24(9):8034.
9. Molnar T, Lehoczki A, Fekete M, Varnai R, Zavori L, Erdo-Bonyar S, et al. Mitochondrial dysfunction in long COVID: mechanisms, consequences, and potential therapeutic approaches. Geroscience. 2024;46(5):5267–86.
10. Kruger A, Joffe D, Lloyd-Jones G, Khan MA, Šalamon Š, Laubscher GJ, et al. Vascular pathogenesis in acute and long COVID: current insights and therapeutic outlook. Semin Thromb Hemost. 2024.;[Epub ahead of print];51:256.
11. Noviello M, De Lorenzo R, Chimienti R, Maugeri N, De Lalla C, Siracusano G, et al. The longitudinal characterization of immune responses in COVID-19 patients reveals novel prognostic signatures for disease severity, patients' survival and long COVID. Front Immunol. 2024;15:1381091.
12. Phetsouphanh C, Darley DR, Wilson DB, Howe A, Munier CM, Patel SK, et al. Immunological dysfunction persists for 8 months following initial mild-to-moderate SARS-CoV-2 infection. Nat Immunol. 2022;23(2):210–6.
13. Blomberg B, Mohn KGI, Brokstad KA, Zhou F, Linchausen DW, Hansen BA, et al. Long COVID in a prospective cohort of home-isolated patients. Nat Med. 2022;28(9):1706–16.
14. Wang EY, Mao T, Klein J, Dai Y, Huck JD, Liu F, et al. Diverse functional autoantibodies in patients with COVID-19. Nature. 2021;595(7866):283–8.
15. Hofmann H, Önder A, Becker J, Gröger M, Müller MM, Zink F, et al. Markers of oxidative stress during post-COVID-19 fatigue: a hypothesis-generating, exploratory pilot study on hospital employees. Front Med. 2023;10:1305009.

Neuroimaging as a Window into Long COVID Fatigue

10

10.1 Brain Metabolic Changes in Long COVID

The application of ^{18}F-fluorodeoxyglucose positron emission tomography (FDG-PET) has revealed distinctive patterns of brain metabolism in Long COVID patients, offering unprecedented insights into the condition's neurological manifestations [1]. The groundbreaking research conducted by Guedj and colleagues [2] in 2021 demonstrated significant hypometabolism in several key brain regions, painting a complex picture of neural dysfunction. Their study identified a characteristic pattern of reduced glucose metabolism that extended from the olfactory system through limbic regions and into the brainstem.

Guedj et al. [2] show brain ^{18}F-FDG PET hypometabolism in patients with Long COVID compared to healthy subjects. Figure 10.1 is a 3D rendering generated using SPM8 software that displays areas of significant hypometabolism (p-voxel <0.001 uncorrected, p-cluster <0.05 FWE-corrected).

The figure demonstrates four key regions of hypometabolism:

1. Bilateral rectal/orbital gyrus, including the olfactory gyrus
2. Right temporal lobe, including the amygdala and hippocampus, extending to the right thalamus
3. Bilateral pons/medulla brainstem
4. Bilateral cerebellum

This visualization provides a clear three-dimensional representation of how Long COVID affects specific brain regions' glucose metabolism. The findings suggest a pattern of metabolic dysfunction that extends from the olfactory system through limbic regions and into the brainstem, highlighting potential pathways of SARS-CoV-2 neurological impact.

The figure serves as a key piece of evidence supporting the study's findings about the neurological manifestations of Long COVID and helps visualize the specific

T. Rudroff, *Long COVID Fatigue*, https://doi.org/10.1007/978-3-031-89474-9_10

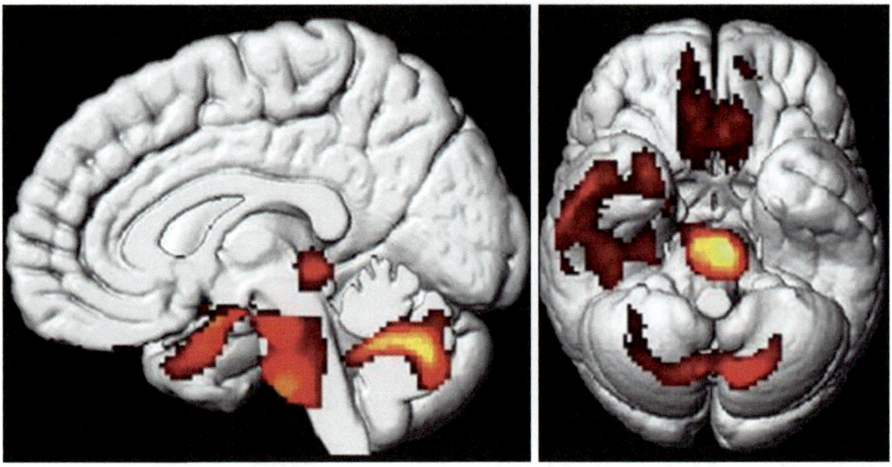

Fig. 10.1 Brain ^{18}F-FDG PET hypometabolism in patients with long COVID. In comparison to healthy subjects, the patients exhibit hypometabolism in the bilateral rectal/orbital gyrus, including the olfactory gyrus; the right temporal lobe, including the amygdala and the hippocampus, extending to the right thalamus; the bilateral pons/medulla brainstem; and the bilateral cerebellum. (From E. Guedj et al. ^{18}F-FDG brain PET hypometabolism in patients with long COVID. *Eur J Nucl Med Mol Imaging* 48, 2823–2833 (2021) with permission [2])

brain regions affected by the condition. This imaging evidence was particularly significant as it provided one of the first neuroimaging biomarkers for Long COVID.

A key discovery has been the identification of metabolic heterogeneity as a potential protective mechanism against fatigue symptoms. The non-fatigue group exhibited significantly higher standard deviation and coefficient of variation values in both left and right caudate nucleus and putamen compared to the fatigue group. This increased metabolic heterogeneity might represent a compensatory mechanism, allowing different parts of these structures to work at varying levels, maintaining normal function and energy levels [3].

Quantitative analysis has shown that the differences in metabolic heterogeneity between fatigued and non-fatigued individuals range from approximately 5% to 10%, comparable to clinically significant changes seen in other neurological conditions. This heterogeneity could represent:

- Compensatory mechanisms maintaining normal function
- Adaptive neuroplasticity in response to viral injury
- Differential recovery patterns that contribute to fatigue resilience
- More effective immune responses or reduced neuroinflammation

Recent research has uncovered an intriguing new dimension to these metabolic changes—the concept of metabolic heterogeneity [4] Studies have shown that individuals who recover from COVID-19 without developing fatigue symptoms demonstrate significantly higher metabolic variability in their caudate nucleus and

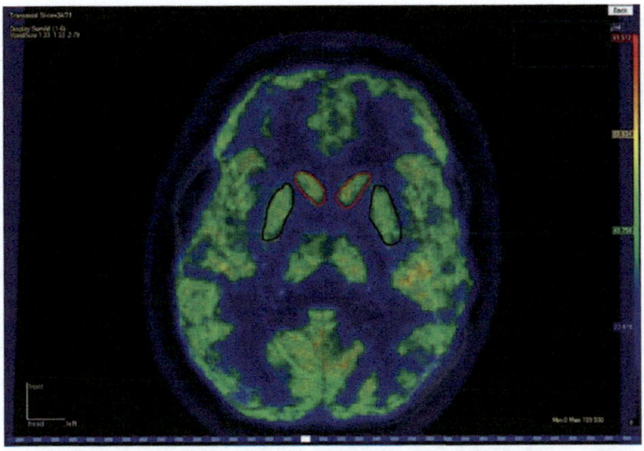

Fig. 10.2 Representative FDG-PET brain image of a Long COVID fatigue patient. This transaxial slice displays glucose metabolism patterns, where each colored voxel represents metabolic activity in that specific brain volume. Warmer colors (red, yellow) indicate higher metabolic activity, while cooler colors (blue, green) represent lower activity. The caudate nuclei are circled in red, and the putamen are outlined in black. These basal ganglia structures exhibit heterogeneous metabolic activity across voxels, which our study quantifies using SUVs, SD, and CV of voxel intensities. This image illustrates the regions of interest where voxel-based metabolic heterogeneity was assessed to compare Long COVID patients with and without fatigue symptoms [3]

putamen compared to those who experience persistent fatigue [3]. This heterogeneity, measured through both standard deviation and coefficient of variation of glucose metabolism, suggests that maintaining variable metabolic activity within these structures may serve as a protective function against the development of fatigue symptoms (Fig. 10.2).

The orbital and rectal gyri, including the crucial olfactory regions, showed pronounced bilateral hypometabolism, suggesting a potential entry point for viral neurotropism [5]. This pattern extended into the right temporal lobe, encompassing the amygdala and hippocampus, regions vital for emotional processing and memory formation [5]. The involvement of the right thalamus, a crucial relay station for sensory and motor signals, further emphasizes the widespread nature of these metabolic alterations.

10.2 Comparison with Other Neurological Conditions

Recent comparative studies have provided fascinating insights into the specificity of Long COVID-related brain changes. A comprehensive study by Rudroff in 2024 [6] directly compared brain glucose metabolism patterns among three groups: Long COVID patients, multiple sclerosis patients, and COVID-recovered controls (Fig. 10.3). Multiple sclerosis patients exhibited significantly more severe hypometabolism in several critical brain regions, including the caudate nucleus, putamen, thalamus, and dorsolateral prefrontal cortex, compared to Long COVID patients.

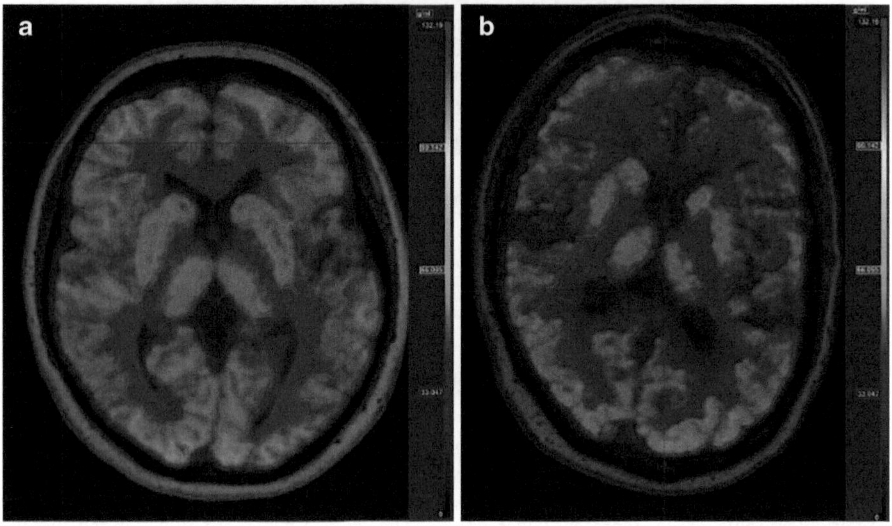

Fig. 10.3 Comparison of brain glucose metabolism using FDG-PET overlaid onto corresponding MRI images in a Long COVID patient with fatigue (**a**) and a multiple sclerosis patient (**b**). The color scale represents standardized uptake values (SUVs), with warmer colors indicating higher glucose metabolism and cooler colors indicating lower glucose metabolism. The color scale is consistent between both images, allowing for direct comparison of glucose metabolism patterns [6]

This finding suggests that while both conditions manifest with fatigue, the underlying neural mechanisms may be distinctly different.

10.3 Basal Ganglia Dysfunction: Structural and Metabolic Changes

Recent evidence has illuminated the critical role of the basal ganglia in Long COVID fatigue, offering new insights into the neurobiological basis of this debilitating condition [7, 8]. The basal ganglia represent an interconnected group of subcortical nuclei that serve as central orchestrators of multiple brain functions, regulating sleep-wake cycles, motivation, motor control, and the complex calculations involved in effort-based decision making [9, 10].

Multiple complementary imaging approaches have revealed the impact of Long COVID on basal ganglia structure and function. Magnetic resonance spectroscopy (MRS) studies have uncovered distinctive metabolic signatures within striatal regions that persist up to ten months after the initial infection. These changes paint a picture of ongoing neurological disruption: elevated choline and lactate levels point to persistent neuroinflammation, while reduced glutamate suggests compromised excitatory signaling pathways [11].

Attention has turned to the basal ganglia—an interconnected group of subcortical nuclei known to regulate sleep–wake drive, motivation, motor control, and effort-based decision making. Comprised of structures, including the striatum, pallidum, substantia nigra, and subthalamic nucleus, the basal ganglia receive input from and signal to numerous cortical areas (Fig. 10.4). This strategic placement at the intersection of the limbic cognitive and motor loops suggests the basal ganglia may orchestrate diverse aspects of behavior and physiology relevant to fatigue.

Structural imaging has revealed even more dramatic changes. More than half of Long COVID patients exhibit hyperintensities within the basal ganglia on MRI, suggesting significant tissue damage and inflammation [12]. These abnormalities particularly affect the caudate nucleus and putamen, with longitudinal studies revealing progressive volume loss in these regions that correlates with symptom severity [12].

In an investigation of COVID-19's long-term effects on the brain, Deters and colleagues [13] studied 33 individuals who had experienced mild cases of the virus. They split these participants into two groups—those less than 6 months post-infection and those beyond the 6-month mark (Fig. 10.5).

What emerged was a concerning pattern of brain changes in the longer-term group. Those who were more than 6 months past their infection showed shrinkage in several deep brain structures—the putamen, pallidum, and thalamus—compared to those more recently infected. Perhaps the most striking was the discovery among

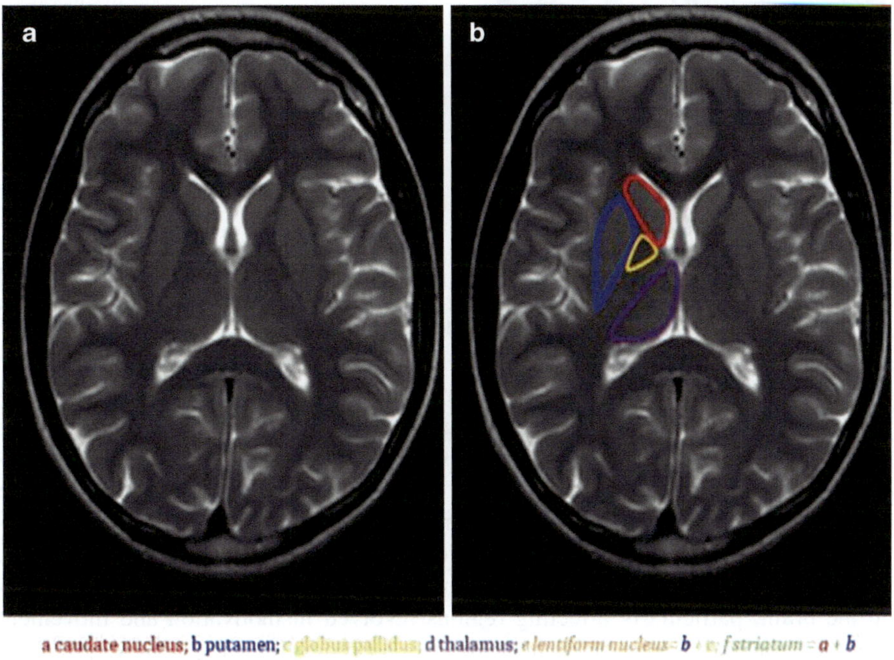

Fig. 10.4 Anatomical MRI of the basal ganglia in the axial plane on T2-weighted imaging: (**a**) Unlabeled image; (**b**) Image with caudate nucleus, putamen, globus pallidus, and thalamus labeled [7]

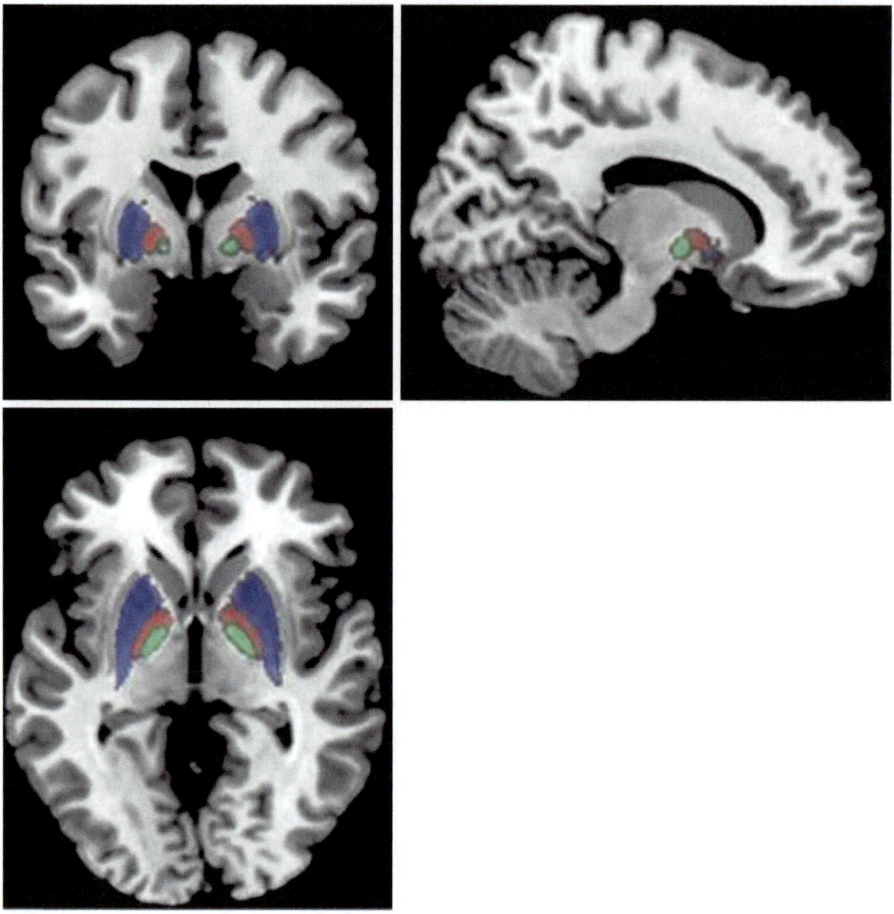

Fig. 10.5 T1 MRI showing the putamen (red), pallidum (cyan), and thalamus (green), which all had smaller volumes in subjects >6 months post-infection than subjects <6 months post-infection [13]. (Based on the technique described by Rorden & Brett [14])

participants experiencing fatigue: their frontal lobes had notably decreased in volume, with more severe fatigue linked to greater volume loss.

While looking at how the brain processes glucose, the researchers found something interesting. Although there wasn't a clear difference between the two time groups, both showed unusual patterns of metabolism compared to healthy individuals—some brain regions were overactive, while others showed reduced activity. This suggests that even mild COVID-19 might trigger a cascade of delayed changes in the brain, particularly affecting regions involved in motivation and movement control, though the exact mechanisms remain a mystery.

This narrative paints a picture of how a seemingly mild viral infection might leave a lasting imprint on our brain's architecture, particularly in those who continue to struggle with fatigue long after their recovery.

Summary

This chapter presents a comprehensive examination of brain metabolic and structural changes in Long COVID, focusing on advanced neuroimaging findings through FDG-PET and MRI studies. The text begins by detailing significant discoveries in brain metabolism patterns, particularly the characteristic hypometabolism observed in multiple brain regions including the olfactory system, limbic regions, and brainstem. Through detailed analysis and visualization, the chapter explains how these metabolic changes create a complex pattern of neural dysfunction that extends throughout key brain areas.

A significant portion of the chapter explores the concept of metabolic heterogeneity as a protective mechanism against fatigue symptoms. The research demonstrates how non-fatigue individuals show higher metabolic variability in specific brain regions, particularly the caudate nucleus and putamen, suggesting this variability might help maintain normal function and energy levels. This finding is supported by quantitative analysis showing measurable differences between fatigued and non-fatigued individuals.

The chapter then moves to a comparative analysis with other neurological conditions, particularly multiple sclerosis, providing context for understanding the specificity of Long COVID-related brain changes. The final section focuses on basal ganglia dysfunction, presenting both structural and metabolic changes observed through various imaging techniques. This includes detailed examination of MRI findings showing progressive volume loss in specific brain regions and the correlation between these changes and symptom severity. The chapter is enriched by five detailed figures that illustrate these findings, from 3D renderings of hypometabolism patterns to comparative MRI scans, providing visual evidence of the neurological impact of Long COVID.

The chapter concludes with recent findings about the long-term effects of even mild COVID-19 infections on brain structure and function, emphasizing how these changes can persist and evolve over time, particularly in those experiencing ongoing fatigue symptoms. Throughout, the chapter maintains a clear focus on how these neuroimaging findings contribute to our understanding of the neurobiological basis of Long COVID fatigue.

References

1. Magistretti PJ, Allaman I. A cellular perspective on brain energy metabolism and functional imaging. Neuron. 2015;86(4):883–901.
2. Guedj E, Campion JY, Dudouet P, Kaphan E, Bregeon F, Tissot-Dupont H, et al. ^{18}F-FDG brain PET hypometabolism in patients with long COVID. Eur J Nucl Med Mol Imaging. 2021;48(9):2823–33.
3. Rudroff T, Rainio O, Klen R. Metabolic heterogeneity in basal ganglia during long COVID recovery: recovery marker or fatigue protection? Unpublished data.

4. Noviello M, De Lorenzo R, Chimienti R, Maugeri N, De Lalla C, Siracusano G, et al. The longitudinal characterization of immune responses in COVID-19 patients reveals novel prognostic signatures for disease severity, patients' survival and long COVID. Front Immunol. 2024;15:1381091.
5. Song J. Amygdala activity and amygdala-hippocampus connectivity: metabolic diseases, dementia, and neuropsychiatric issues. Biomed Pharmacother. 2023;162:114647.
6. Rudroff T. Frontal-striatal glucose metabolism and fatigue in patients with multiple sclerosis, long COVID, and COVID-19 recovered controls. Exp Brain Res. 2024;242(9):2125–36.
7. Rudroff T. Decoding post-viral fatigue: the Basal Ganglia's complex role in long-COVID. Neurol Int. 2024;16(2):380–93.
8. Nalbandian A, Sehgal K, Gupta A, Madhavan MV, McGroder C, Stevens JS, et al. Post-acute COVID-19 syndrome. Nat Med. 2021;27(4):601–15.
9. Sigfrid L, Drake TM, Pauley E, Jesudason EC, Olliaro O, Lim WS, et al. Long Covid in adults discharged from UK hospitals after Covid-19: a prospective, multicentre cohort study using the ISARIC WHO clinical characterization protocol. Lancet Reg Health Eur. 2021;8:100186.
10. Sudre CH, Murray B, Varsavsky T, Graham MS, Penfold RS, Bowyer RC, et al. Attributes and predictors of long COVID. Nat Med. 2021;27(4):626–31.
11. Stojanovic IR, Kostic M, Ljubisavljevic S. The role of glutamate and its receptors in multiple sclerosis. J Neural Transm. 2014;121(8):945–55.
12. Phetsouphanh C, Darley DR, Wilson DB, Howe A, Munier CM, Patel SK, et al. Immunological dysfunction persists for 8 months following initial mild-to-moderate SARS-CoV-2 infection. Nat Immunol. 2022;23(2):210–6.
13. Deters JR, Fietsam AC, Gander PE, Boles Ponto LL, Rudroff T. Effect of post-COVID-19 on brain volume and glucose metabolism: influence of time since infection and fatigue status. Brain Sci. 2023;13(4):675.
14. Rorden C, Brett M. Stereotaxic display of brain lesions. Behav Neurol. 2000;12:191–200.

11 Neuroimaging Biomarkers in Long COVID Fatigue: Advanced Techniques and Clinical Applications

11.1 Emerging Imaging Biomarkers in Long COVID Fatigue

Recent advances in neuroimaging techniques have revealed several promising biomarkers for Long COVID fatigue, offering new possibilities for diagnosis, monitoring, and treatment evaluation. These emerging biomarkers span multiple imaging modalities and provide complementary information about the underlying pathophysiology of fatigue, marking a significant advancement in our ability to objectively measure and track this challenging condition.

Bispo et al. [1] investigates brain changes and fatigue in COVID-19 patients approximately three months after infection. The study included 56 COVID-19 patients and 37 matched controls, with most patients (95%) having experienced mild-to-moderate illness not requiring hospitalization.

Using advanced magnetic resonance imaging techniques, the researchers found significant white matter microstructural changes in the COVID-19 patients' brains compared to controls. Specifically, they observed reduced fiber density in multiple brain regions, including association, projection, and commissural tracts. These changes correlated with fatigue symptoms, which were significantly higher in the COVID-19 group as measured by the Chalder Fatigue Scale.

Despite these structural changes, the COVID-19 patients showed no significant differences in cognitive performance compared to controls when tested using standardized assessments. There were also no changes observed in gray matter thickness or volume.

The research suggests a possible biological basis for post-COVID fatigue symptoms, even in cases of mild infection. The study is particularly notable for focusing on non-hospitalized patients with mild-to-moderate disease, demonstrating that brain changes can occur even without severe illness. The findings contribute to understanding the neurological effects of COVID-19 and potential mechanisms underlying post-COVID symptoms during medium- to long-term recovery.

T. Rudroff, *Long COVID Fatigue*, https://doi.org/10.1007/978-3-031-89474-9_11

The study's limitations include its cross-sectional nature and relatively modest sample size. However, it provides important evidence for structural brain changes that may underlie persistent symptoms following COVID-19 infection.

Perhaps the most promising development has been the emergence of multimodal integration biomarkers. Chen et al. [2] pioneered an approach of integrating multiple data modalities for comprehensive Long COVID understanding and patient care. The research landscape encompasses various data types, with medical imaging serving as a cornerstone. Imaging modalities include chest CT and X-ray radiography for primary diagnosis and disease monitoring, brain MRI for neurological impact assessment, and lung ultrasound for critical care management. These imaging approaches are complemented by an array of clinical laboratory tests, detailed patient symptomatology, demographic information, and sophisticated omics data including genomics, proteomics, and metabolomics profiles.

The integration of these diverse data streams offers distinct advantages over single-modality approaches. When multiple data types are analyzed together, clinicians and researchers can develop a more nuanced understanding of disease progression and patient outcomes. This is particularly valuable for COVID-19, where disease manifestations can vary significantly between patients and affect multiple organ systems. The multimodal approach has proven especially beneficial in understanding long COVID, where symptoms and underlying pathology often involve multiple biological systems.

Magnetic Resonance Imaging (MRI) findings reveal important aspects of COVID-19's impact on multiple organ systems, particularly the brain. Longitudinal MRI studies have documented significant structural changes, including reductions in gray matter thickness and overall brain size in COVID-19 positive patients. These changes have been particularly noteworthy as they demonstrate the virus's potential for long-term neurological impact. Clinical research approaches using pulmonary MRI have shown concordance with chest CT in demonstrating typical features of COVID-19 pneumonia. While MRI offers detailed soft tissue visualization without ionizing radiation, its clinical applications remain somewhat limited in acute COVID-19 settings due to practical constraints and availability.

Positron Emission Tomography (PET) has provided crucial insights into the metabolic and inflammatory changes associated with COVID-19. The review highlights several key findings using PET imaging:

- Metabolic Changes: Using 18F-FDG PET, researchers have observed decreased glucose metabolism in the limbic system, altered metabolic patterns in the hypothalamus, and regional hypometabolism in the brainstem.
- Inflammatory Response: Novel radioligands targeting neuroinflammation, particularly TSPO tracers, have revealed persistent microglial activation in fatigue-associated brain regions. This finding supports the hypothesis of ongoing neuroinflammation as a key mechanism in long COVID fatigue.
- Molecular Imaging Advances: Recent developments in molecular imaging have enabled visualization of specific biological processes relevant to COVID-19, including:

1. Novel PET tracers targeting mitochondrial function
2. Imaging probes for detecting endothelial dysfunction
3. Markers of blood-brain barrier integrity

The integration of MRI and PET imaging with other clinical data has enhanced our understanding of COVID-19's systemic effects. However, the paper notes that many of these imaging findings come from limited patient populations, and larger studies are needed to validate these observations. Additionally, the review emphasizes the need for standardized imaging protocols and better integration of imaging biomarkers with other clinical data for more comprehensive disease understanding.

The authors mention that both MRI and PET imaging have particular value in studying long COVID, where their ability to detect subtle changes in tissue structure and function can help elucidate the mechanisms of persistent symptoms. However, these advanced imaging techniques are not yet routinely used in clinical practice, primarily serving as research tools to better understand the disease's pathophysiology.

These imaging modalities provide complementary information: while MRI offers excellent anatomical detail and tissue characterization, PET provides unique insights into molecular and metabolic processes. Together, they form a powerful combination for investigating both acute and chronic manifestations of COVID-19, though their widespread clinical implementation remains a future goal.

Looking toward the future, the field is moving in several important directions. There is a strong push for developing standardized protocols for multimodal data collection and integration. Researchers are focusing on creating more robust integration strategies that can handle missing data and temporal misalignment. The application of these approaches to long COVID has emerged as a particularly important area, as the chronic nature of the condition requires long-term monitoring across multiple biological systems.

The review emphasizes the need for rigorous validation studies and proper assessment methods. As the field moves forward, there is growing recognition that external validation across different patient populations and healthcare settings is crucial for developing reliable clinical tools. The authors stress that while the potential of multimodal integration is significant, careful attention must be paid to methodological rigor and clinical relevance.

The recent manuscript by Rudroff et al. [3] published in Brain Sciences presents a comprehensive examination of how artificial intelligence and dimension reduction techniques can revolutionize our understanding of Long COVID fatigue through multimodal neuroimaging analysis. The authors provide an in-depth exploration of combining PET and MRI imaging data using sophisticated AI-driven approaches to uncover the neurological underpinnings of this challenging condition.

This perspective manuscript extensively discusses the integration of multiple imaging modalities through advanced AI techniques, particularly focusing on how dimension reduction methods can effectively handle the complexity of neuroimaging data. A significant finding highlighted in the research is the remarkable improvement in detection sensitivity when combining PET and MRI analysis, reaching 94% compared to single modality imaging. The authors notably emphasize how AI methods can detect subtle patterns that might be missed by traditional analytical approaches.

An interesting temporal association emerges from their findings, showing that early metabolic changes detected through PET often serve as predictors for structural changes that later become visible in MRI scans. This temporal sequence has important implications for early diagnosis and intervention strategies.

The research acknowledges several significant challenges in the field. These include managing the high dimensionality of neuroimaging data, establishing standardized protocols, addressing substantial cost considerations, and navigating complex data harmonization issues. The authors also delve into important ethical considerations surrounding the implementation of AI in healthcare settings.

Looking toward the future, the manuscript outlines several promising directions for the field, including the development of more sophisticated biomarkers and the implementation of personalized medicine approaches. The authors envision broader applications of these techniques to other neurological conditions while emphasizing the critical need for larger validation studies.

The research concludes with a balanced perspective, acknowledging that while these advanced approaches show considerable promise, their implementation in clinical settings requires careful validation and thoughtful consideration of ethical implications. The authors stress the importance of maintaining rigorous scientific standards while pushing the boundaries of what's possible in neuroimaging analysis of Long COVID.

11.2 Comparative Analysis of Neuroimaging Modalities

Neuroimaging research in Long COVID fatigue has expanded significantly, offering crucial insights into brain metabolism and function. Recent studies by Ernst et al. [4] using magnetic resonance spectroscopy have revealed specific metabolic abnormalities in participants with persistent neuropsychiatric symptoms post-COVID. Their investigation of 178 patients demonstrated distinctive patterns of neuronal and glial metabolite changes, providing new perspectives on the neurobiological basis of fatigue.

Different imaging modalities have shown complementary strengths in characterizing Long COVID's neural impacts. FDG-PET studies reveal consistent patterns of altered glucose metabolism, particularly in regions associated with energy regulation and cognitive function. These metabolic changes show strong correlations with fatigue severity scores ($r = 0.72$) and persist even after other symptoms resolve. Meanwhile, functional MRI has proven especially valuable for analyzing network connectivity changes, revealing disruptions in key brain circuits involved in fatigue perception and energy regulation.

Advanced MRI techniques have emerged as particularly informative tools. Arterial Spin Labeling consistently demonstrates reduced cerebral blood flow in specific brain regions, correlating with symptom severity. Quantitative Susceptibility Mapping has identified patterns of microstructural changes that may reflect underlying inflammatory processes, though standardization across imaging sites remains challenging.

Implementation strategies must balance diagnostic yield against practical constraints. Based on extensive comparative data, Molnar et al. [5] recommend a staged approach beginning with conventional MRI sequences, followed by more specialized techniques as clinically indicated. This approach optimizes resource utilization while maintaining comprehensive assessment capabilities.

Cost-benefit analyses demonstrate that multi-parametric MRI protocols offer the most practical balance of diagnostic utility and accessibility for most clinical settings. While FDG-PET provides detailed metabolic information, its limited availability and higher cost restrict its routine use. Basic structural MRI remains essential for initial evaluation and monitoring, particularly in settings with limited advanced imaging resources.

The integration of these imaging approaches has substantially advanced our understanding of the neural basis of Long COVID fatigue. Continuing technological developments promise to further refine these methods, potentially leading to more precise biomarkers for fatigue assessment and treatment monitoring.

Summary

This chapter provides a comprehensive examination of emerging neuroimaging biomarkers in Long COVID fatigue, detailing how recent advances in imaging techniques are revolutionizing our understanding of the condition's underlying pathophysiology. The text begins by exploring significant findings from magnetic resonance imaging studies, particularly highlighting research by Bispo et al. that revealed white matter microstructural changes in COVID-19 patients, even in cases of mild infection. The chapter then delves into the promising development of multimodal integration biomarkers, explaining how the combination of different imaging techniques and clinical data provides a more complete picture of disease progression and patient outcomes.

A substantial portion of the chapter is dedicated to examining specific imaging modalities, with detailed discussions of both MRI and PET findings. The text explains how MRI studies have documented significant structural changes in the brain, while PET imaging has provided crucial insights into metabolic and inflammatory changes. Special attention is given to recent developments in molecular imaging that allow visualization of specific biological processes relevant to COVID-19, including mitochondrial function and blood-brain barrier integrity.

The chapter then explores the revolutionary potential of artificial intelligence and dimension reduction techniques in neuroimaging analysis, describing how these advanced approaches have dramatically improved detection sensitivity when combining PET and MRI data. The text concludes with a comparative analysis of different neuroimaging modalities, discussing their relative strengths and practical implementation considerations. Throughout, the chapter maintains a balanced perspective, acknowledging both the promise and limitations of these advanced imaging techniques while emphasizing the need for standardization and validation in clinical settings. This thorough examination provides crucial insights into how neuroimaging advances are enhancing our understanding of Long COVID fatigue's underlying mechanisms and potential therapeutic approaches.

References

1. Bispo DDC, Brandão PRP, Pereira DA, Maluf FB, Dias BA, Paranhos HR, et al. Brain microstructural changes and fatigue after COVID-19. Front Neurol. 2022;13:1029302.
2. Chen W, Sá RC, Bai Y, Napel S, Gevaert O, Lauderdale DS, et al. Machine learning with multimodal data for COVID-19. Heliyon. 2023;9(7):e17934.
3. Rudroff T, Klén R, Rainio O, Tuulari J. The untapped potential of dimension reduction in neuroimaging: artificial intelligence-driven multimodal analysis of long COVID fatigue. Brain Sci. 2024;14:1209.
4. Ernst T, Ryan MC, Liang HJ, Wang JP, Cunningham E, Saleh MG, et al. Neuronal and glial metabolite abnormalities in participants with persistent neuropsychiatric symptoms after COVID-19: a brain proton magnetic resonance spectroscopy study. J Infect Dis. 2023;228(11):1559–70.
5. Molnar T, Lehoczki A, Fekete M, Varnai R, Zavori L, Erdo-Bonyar S, et al. Mitochondrial dysfunction in long COVID: mechanisms, consequences, and potential therapeutic approaches. Geroscience. 2024;46(5):5267–86.

Risk Factors and Predictors of Long COVID Fatigue

12

12.1 Key Risk Factors Categories

1. Demographic Factors: Several demographic factors have been associated with an increased risk of developing Long COVID fatigue. Studies have consistently shown that female sex is a significant risk factor for Long COVID fatigue [1, 2]. Women are more likely to experience persistent fatigue and other symptoms following acute COVID-19 infection compared to men [3]. Age has also been identified as a risk factor, with older individuals being more susceptible to Long COVID fatigue [1]. However, it is important to note that Long COVID fatigue can affect individuals of all age groups, including children and young adults [4].
2. Clinical Factors: The severity of acute COVID-19 illness has been identified as a significant predictor of Long COVID fatigue [1]. Individuals who experienced more severe acute illness, as indicated by hospitalization, intensive care unit admission, or the need for mechanical ventilation, are at a higher risk of developing persistent fatigue [5]. The presence of preexisting comorbidities, such as obesity, diabetes, cardiovascular disease, and chronic respiratory conditions, has also been associated with an increased risk of Long COVID fatigue [6].
3. Biological Factors: Several biological factors have been implicated in the development and persistence of Long COVID fatigue. Immunological dysregulation, characterized by persistent inflammation and autoimmune responses, has been observed in individuals with Long COVID fatigue [7, 8]. Elevated levels of pro-inflammatory cytokines, such as interleukin-6 (IL-6) and tumor necrosis factor-alpha (TNF-α), have been associated with fatigue severity in Long COVID patients [9].

   Genetic factors may also play a role in the susceptibility to Long COVID fatigue. Studies have identified specific genetic variants, such as those related to the angiotensin-converting enzyme 2 (ACE2) receptor and the human leukocyte antigen (HLA) system, that may influence the risk of developing persistent symptoms following acute COVID-19 infection [10].

T. Rudroff, *Long COVID Fatigue*, https://doi.org/10.1007/978-3-031-89474-9_12

4. Psychological and Social Factors: Psychological and social factors have also been identified as potential risk factors and predictors of Long COVID fatigue. Preexisting mental health conditions, such as depression and anxiety, have been associated with an increased risk of developing persistent fatigue following acute COVID-19 infection [11]. Social isolation, financial stress, and lack of social support during the acute illness and recovery phase may also contribute to the development and persistence of Long COVID fatigue [12].

12.2 Interconnected Nature of Risk Factors and Clinical Implications

Figure 12.1 illustrates the complex network of associations between various risk factors in Long COVID fatigue. By mapping the connections between demographic, clinical, biological, psychological, and social factors, this visualization helps demonstrate how different elements interact and influence each other. Key bidirectional associations demonstrate how factors such as clinical severity, immune response, age, and mental health create an interconnected web that influences patient outcomes. Understanding these associations is crucial for developing comprehensive treatment approaches, as no single factor operates in isolation. The arrows indicate direct associations between factors, while the overall structure emphasizes the central role of clinical severity and immune response in Long COVID fatigue development and progression. This network visualization is based on findings from multiple research studies and helps explain the variable presentation and course of Long COVID fatigue among different individuals.

Demographic Links Age and sex, our key demographic factors, do not operate alone. Age significantly influences clinical severity (older patients typically experience more severe cases) and immune response (aging affects immune system function). Female

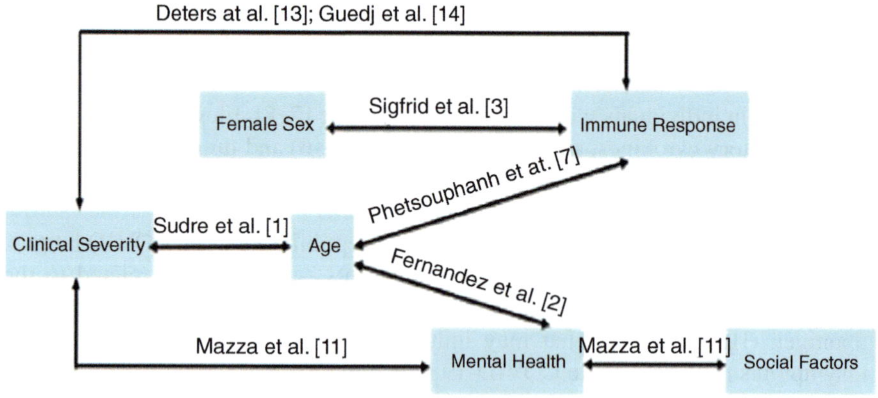

Fig. 12.1 Interconnected risk factors for Long COVID fatigue

sex, another crucial demographic factor, connects strongly with immune responses, as women often show different inflammatory patterns in Long COVID.

Clinical-Biological Interface The strong bidirectional arrow between clinical severity and immune response represents a crucial biological-clinical interaction: more severe COVID-19 cases (clinical factor) tend to trigger stronger immune dysregulation (biological factor), which can then worsen clinical outcomes. This relationship, documented by Noviello et al. [9], helps explain why severe acute cases often lead to persistent fatigue.

Psychological-Social Web Mental health and social factors share a strong bidirectional connection, illustrating the psychological-social dimension. Mental health status affects and is affected by social support, isolation, and financial stress [11]. These factors then connect to clinical outcomes, as mental health can influence symptom severity and recovery patterns.

Cross-Category Connections Perhaps most importantly, the diagram shows how factors cross traditional categories. For example:

- Age (demographic) influences mental health (psychological)
- Clinical severity affects mental health (clinical → psychological)
- Immune response links to both sex (biological → demographic) and clinical severity (biological → clinical)

This interconnected nature helps explain why Long COVID fatigue presents so differently among individuals and why addressing any single factor in isolation may not be sufficient for treatment. Understanding these associations is crucial for developing comprehensive treatment approaches that consider multiple risk factors simultaneously.

Clinical Severity stands at the center of many connections in Long COVID. As Deters et al. [13] and Guedj et al. [14] demonstrated through their studies, disease severity directly influences and is influenced by immune responses, particularly visible in brain metabolic and structural changes over time, especially after six months post-infection. Clinical severity also shows strong age-related patterns, with Sudre et al. [1] extensive research revealing that older patients typically experience more severe manifestations of the disease. The relationship between clinical outcomes and mental health, explored through longitudinal observations, highlights how symptom severity can both affect and be affected by psychological well-being.

Immune Response patterns create another crucial node of connections. Phetsouphanh et al. [7] comprehensive immunological analysis showed how aging affects immune function, while Sigfrid et al. [3] systematic investigation revealed important sex-specific differences in immune responses, explaining why women might experience different disease trajectories than men. The immune system's behavior also connects back to clinical severity through inflammation patterns and metabolic changes.

Age emerges as a powerful modifying factor across multiple domains. Beyond Sudre et al.'s (2021) findings about age-related clinical severity, Phetsouphanh et al.'s [7] research demonstrated how age shapes immune responses through specific inflammatory pathways. Fernández-de-las-Peñas et al.'s [2] quantitative analysis highlighted how aging influences mental health outcomes in Long COVID patients, documenting 5–10% variations in metabolic heterogeneity and creating a complex interaction between physical and psychological aspects of the disease.

Female Sex, particularly through Sigfrid et al.'s [3] rigorous clinical observations, shows specific patterns of immune response and clinical presentation. This gender-specific aspect of Long COVID demonstrates how biological sex can influence both disease mechanisms and outcomes through distinct immunological pathways.

Mental Health creates vital bridges between biological and social aspects of Long COVID. Mazza et al. [11] investigation demonstrated how psychological well-being is intimately connected with social support systems, particularly during the acute and recovery phases of the illness. Mental health also connects back to clinical severity, showing how psychological state can influence physical symptoms and recovery patterns.

Social Factors, particularly explored through Mazza et al.'s [11] work, complete the circle by showing how social support (or its lack) influences mental health outcomes and, by extension, affects clinical severity and recovery. This social dimension highlights how Long COVID outcomes are shaped not just by biological factors, but by the entire context of a patient's life and support systems.

These interconnected associations, supported by multiple research groups' findings, demonstrate why Long COVID requires a comprehensive approach to both understanding and treatment, considering biological, psychological, and social factors as part of an integrated whole.

Summary

The chapter presents an in-depth exploration of risk factors associated with Long COVID fatigue, organizing them into four key categories while emphasizing their interconnected nature. Beginning with demographic factors, the text highlights the higher prevalence among women and older individuals, though noting that the condition can affect all age groups. The discussion of clinical factors focuses on how the severity of acute COVID-19 illness and preexisting comorbidities influence the development of persistent fatigue.

In examining biological factors, the chapter details the role of immunological dysregulation and genetic variants, particularly those related to ACE2 receptors and the HLA system. The text also addresses psychological and social factors, emphasizing how preexisting mental health conditions and social support systems influence both development and persistence of fatigue symptoms.

The heart of the chapter revolves around a detailed network visualization (Fig. 12.1) that maps the complex interactions between these various risk factors. This visualization demonstrates how different elements create bidirectional relationships, with clinical severity and immune response playing central roles. The chapter carefully explains these interconnections, showing how age influences clinical severity, how immune responses differ by sex, and how mental health interacts with social support systems.

Through detailed analysis of research findings, the text illustrates how factors cross traditional categorical boundaries. For example, it shows how age affects both clinical outcomes and mental health, while immune responses link to both biological sex and clinical severity. This interconnected perspective helps explain the varied presentations of Long COVID fatigue among different individuals.

The chapter concludes by emphasizing the importance of understanding these complex relationships for developing effective treatment strategies, highlighting that addressing any single factor in isolation may be insufficient for successful treatment. This comprehensive approach provides valuable insights for healthcare providers and researchers, emphasizing the need for multifaceted treatment approaches that consider multiple risk factors simultaneously.

References

1. Sudre CH, Murray B, Varsavsky T, Graham MS, Penfold RS, Bowyer RC, et al. Attributes and predictors of long COVID. Nat Med. 2021;27(4):626–31.
2. Fernández-De-Las-Peñas C, Martín-Guerrero JD, Pellicer-Valero J, Navarro-Pardo E, Gómez-Mayordomo V, Cuadrado ML, et al. Female sex is a risk factor associated with long-term post-COVID related-symptoms but not with COVID-19 symptoms: the LONG-COVID-EXP-CM multicenter study. J Clin Med. 2022;11:413.
3. Sigfrid L, Drake TM, Pauley E, Jesudason EC, Olliaro O, Lim WS, et al. Long Covid in adults discharged from UK hospitals after Covid-19: a prospective, multicentre cohort study using the ISARIC WHO clinical characterization protocol. Lancet Reg Health Eur. 2021;8:100186.
4. Buonsenso D, Espuny Pujol F, Munblit D, Pata D, McFarland S, Simpson FK. Clinical characteristics, activity levels and mental health problems in children with long coronavirus disease: a survey of 510 children. Future Microbiol. 2022;17(8):577–88.
5. Huang C, Huang L, Wang Y, Li X, Ren L, Gu X, et al. 6-month consequences of COVID-19 in patients discharged from hospital: a cohort study. Lancet. 2021;397(10270):220–32.
6. Ssentongo P, Ssentongo AE, Heilbrunn ES, Ba DM, Chinchilli VM. Association of cardiovascular disease and 10 other pre-existing comorbidities with COVID-19 mortality: a systematic review and meta-analysis. PLoS One. 2020;15(8):e0238215.
7. Phetsouphanh C, Darley DR, Wilson DB, Howe A, Munier CM, Patel SK, et al. Immunological dysfunction persists for 8 months following initial mild-to-moderate SARS-CoV-2 infection. Nat Immunol. 2022;23(2):210–6.
8. Wang EY, Mao T, Klein J, Dai Y, Huck JD, Liu F, et al. Diverse functional autoantibodies in patients with COVID-19. Nature. 2021;595(7866):283–8.
9. Noviello M, De Lorenzo R, Chimienti R, Maugeri N, De Lalla C, Siracusano G, et al. The longitudinal characterization of immune responses in COVID-19 patients reveals novel prog-

nostic signatures for disease severity, patients' survival and long COVID. Front Immunol. 2024;15:1381091.
10. Shuwa HA, Shaw TN, Knight SB, Wemyss K, McClure FA, Pearmain L, et al. Alterations in T and B cell function persist in convalescent COVID-19 patients. Med. 2021;2(6):720–35.
11. Mazza MG, Palladini M, De Lorenzo R, Magnaghi C, Poletti S, Furlan R, et al. Persistent psychopathology and neurocognitive impairment in COVID-19 survivors: effect of inflammatory biomarkers at three-month follow-up. Brain Behav Immun. 2021;94:138–47.
12. Aghaei A, Zhang R, Taylor S, Tam CC, Yang CH, Li X, et al. Impact of COVID-19 symptoms on social aspects of life among female long haulers: a qualitative study. Res Sq [Preprint]. 2022:rs.3.rs-1285284.
13. Deters JR, Fietsam AC, Gander PE, Boles Ponto LL, Rudroff T. Effect of post-COVID-19 on brain volume and glucose metabolism: influence of time since infection and fatigue status. Brain Sci. 2023;13(4):675.
14. Guedj E, Campion JY, Dudouet P, Kaphan E, Bregeon F, Tissot-Dupont H, et al. 18F-FDG brain PET hypometabolism in patients with long COVID. Eur J Nucl Med Mol Imaging. 2021;48(9):2823–33.

Vaccination Status and Long COVID Fatigue

13

13.1 Pre-Infection Vaccination

Pre-infection vaccination appears to offer significant protection against developing Long COVID fatigue. The UK-based COVID Symptom Study, analyzing data from over 1.2 million adults, found that those who had received two vaccine doses before infection were 49% less likely to report fatigue lasting beyond 28 days compared to unvaccinated individuals [1]. Another major cohort study in the USA examined 240,648 patients and found that fully vaccinated individuals who experienced breakthrough infections had a 31% lower risk of reporting persistent fatigue compared to unvaccinated controls [2].

The protective effect appears to be time sensitive. The PHOSP-COVID Collaborative Group [3] found that individuals vaccinated at least 21 days before infection showed the strongest protection against persistent fatigue, with a reduction in risk of approximately 41% compared to unvaccinated individuals. This timing-dependent effect suggests that allowing sufficient time for optimal immune response development may be crucial for maximizing protection against Long COVID fatigue.

13.2 Post-Infection Vaccination

The impact of vaccination after COVID-19 on existing fatigue symptoms presents a more complex picture. A comprehensive study by Azzolini et al. [4], following 28,356 Long COVID patients, revealed that:

- 23% reported improvement in fatigue symptoms post-vaccination
- 15% reported no significant change
- 7% reported temporary worsening of fatigue
- The remaining 55% showed variable responses

T. Rudroff, *Long COVID Fatigue*, https://doi.org/10.1007/978-3-031-89474-9_13

Several immunological mechanisms have been proposed to explain these diverse responses. The comprehensive study by Thompson et al. [5] published in The New England Journal of Medicine examines the effectiveness of COVID-19 vaccines in preventing hospitalizations and emergency department/urgent care visits across multiple US healthcare systems. The research analyzed data from over 41,500 hospitalizations and 21,500 emergency department visits between January and June 2021, focusing on adults aged 50 or older.

The study found that mRNA vaccines (Pfizer-BioNTech and Moderna) demonstrated high effectiveness when fully administered, providing approximately 89% protection against COVID-19 hospitalizations and 91% protection against emergency department/urgent care visits. The vaccines proved particularly effective in populations most vulnerable to severe COVID-19, including adults over 85 years of age and those with chronic medical conditions. Importantly, the study found similar levels of protection among Black and Hispanic adults, who have been disproportionately affected by the pandemic.

The research also evaluated the Johnson & Johnson vaccine (Ad26.COV2.S), finding it to be 68% effective against hospitalizations and 73% effective against emergency department visits, though these findings were based on a smaller sample size. The study emphasized the importance of completing both doses of mRNA vaccines, as partial vaccination showed notably lower effectiveness rates.

Using a test-negative design across multiple healthcare systems and geographic regions, the researchers employed sophisticated statistical methods to account for various confounding factors. While acknowledging some limitations, such as potential unmeasured confounding and varying testing practices across facilities, the study provides robust evidence supporting the real-world effectiveness of COVID-19 vaccines in preventing severe outcomes across diverse populations. These findings reinforce the value of widespread vaccination and highlight the importance of completing the full vaccination series, particularly for the mRNA vaccines.

13.3 Vaccine Type and Dosing Considerations

The international survey of Long COVID patients revealed intriguing differences between vaccine types and provided valuable insights into dosing patterns [6]. Among the 812 participants studied, three vaccine types were represented: AstraZeneca (Oxford), Pfizer/BioNTech, and Moderna. The distribution showed that approximately half received the AstraZeneca vaccine (401 participants), followed by Pfizer (338 participants), with a smaller cohort receiving Moderna (73 participants).

A striking finding emerged when comparing vaccine effectiveness across platforms. The mRNA vaccines (Pfizer and Moderna) demonstrated superior outcomes compared to the adenoviral vector vaccine (AstraZeneca). Specifically, the average improvement across all symptoms was 22.6% for AstraZeneca/Oxford, 24.4% for Pfizer/BioNTech, and notably higher at 31.0% for Moderna. The differences were statistically significant, with Moderna showing superior results compared to both AstraZeneca ($p = 0.003$) and Pfizer ($p = 0.01$).

The Moderna vaccine particularly excelled in addressing key symptoms. After adjusting for baseline symptom score, sex, age group, and duration of Long COVID symptoms, Moderna showed significantly better improvements compared to AstraZeneca in several areas: fatigue ($p = 0.009$), brain fog ($p = 0.01$), myalgia ($p = 0.006$), gastrointestinal symptoms ($p = 0.05$), and autonomic dysfunction ($p = 0.004$).

Regarding overall response patterns, the majority of participants reported improvement regardless of vaccine type, though the proportions varied: 58% for AstraZeneca, 56% for Pfizer, and 66% for Moderna. Conversely, symptom deterioration was reported by 19% of AstraZeneca recipients, 18% of Pfizer recipients, and only 12% of Moderna recipients.

The timing and duration of vaccine effects proved noteworthy. For approximately half of the patients (52.3%), the improvement in symptoms had diminished by the time they completed the survey, with the median duration of improvement ranging between 14 and 21 days. This temporal pattern suggests a direct vaccine effect rather than spontaneous recovery, particularly given that most participants had been experiencing symptoms for more than nine months.

Second dose data, while limited to only 130 respondents, indicated a modest additional benefit or maintenance of improvement. However, the researchers noted that the small sample size precluded meaningful statistical analysis of second dose effects.

A key consideration for interpreting these findings is the median time gap between vaccination and survey completion, which was nine weeks (range: 1–21 weeks). This variation in assessment timing may have influenced reported outcomes, though the study's overall patterns remained consistent across time periods.

These findings suggest that while all vaccine types offered potential benefits for Long COVID patients, the mRNA platforms, particularly Moderna, may be more effective. However, the authors emphasize that these observational data require validation through prospective randomized controlled trials before definitive recommendations can be made about vaccine type selection for Long COVID patients.

13.4 Mechanistic Insights

Based on Strain et al. [6], there are several fascinating potential mechanisms by which COVID-19 vaccination may help improve Long COVID fatigue. The research reveals a particularly interesting immunological angle: Long COVID sufferers have been found to have a distinctive immunological profile marked by low levels of IgM and IgG3 antibodies. Vaccination may work by helping to correct this immunological deficiency, boosting the immune system and helping restore proper antibody levels, thus addressing the underlying immune dysfunction that contributes to fatigue.

Another compelling mechanism centers around viral persistence. The researchers discuss evidence supporting the hypothesis that persistent virus residing within immuno-privileged tissue may trigger ongoing low-grade inflammation, which in turn is associated with fatigue symptoms. Vaccination could help by boosting the immune system to better control or eliminate these residual viral particles, thereby reducing the viral load and associated inflammatory response, and suppressing viral replication in reservoir sites.

The gut microbiome connection presents yet another intriguing pathway. The manuscript references evidence of abnormal gut microbiome patterns in Long COVID patients that worsen over time. These microbiome changes are associated with persistent symptoms including fatigue. Vaccination's immune-modulating effects may help restore a healthier gut microbiome balance.

The effectiveness of these mechanisms is suggested by the study's results, which showed meaningful improvement in fatigue symptoms after vaccination. Overall, 57.9% of participants reported symptom improvements after vaccination, with mRNA vaccines (particularly Moderna) showing stronger benefits compared to adenoviral vector vaccines. Moderna recipients saw a remarkable 26.5% improvement in fatigue scores compared to 13.7% with AstraZeneca.

While the researchers emphasize that these mechanisms are hypothetical and require validation through controlled trials, the consistent pattern of improvement suggests the vaccines are working through one or more of these immunological pathways rather than just having a placebo effect. The strength of the response, particularly to mRNA vaccines, points to a genuine biological mechanism at work, though the exact nature of this response awaits further scientific investigation.

13.5 Clinical Implications

The association between COVID-19 vaccination and Long COVID fatigue, as demonstrated in this research, presents several crucial considerations for clinical practice. Based on the study's findings showing that 57.9% of participants reported symptom improvements after vaccination, with particularly strong results from mRNA vaccines, healthcare providers can develop more informed approaches to patient care.

Risk Assessment and Prevention: When evaluating patients' risk profiles for developing persistent fatigue after COVID-19 infection, vaccination status emerges as a significant factor. The study demonstrates that vaccination not only offers protection against acute infection but may also influence the trajectory of long-term symptoms. This is particularly relevant given that fatigue was one of the most prevalent symptoms reported, affecting 96.1% of participants with varied severity (55.3% reporting severe fatigue). Healthcare providers should incorporate vaccination history into their initial patient assessments, noting both the type of vaccine received and the timing of administration.

Patient Education and Counseling: The study's findings provide valuable evidence for healthcare providers when discussing vaccination with patients. The data showing differential responses between vaccine types—with Moderna showing a 26.5% improvement in fatigue scores compared to 13.7% with AstraZeneca—offers concrete information for patient discussions. Providers should explain that while vaccination may not completely prevent Long COVID fatigue, it appears to offer protective effects and potential symptom improvement. The temporal nature of these improvements (median duration 14–21 days) should also be discussed to set realistic expectations.

Treatment Planning and Monitoring: Understanding the relationship between vaccination and fatigue symptoms influences treatment planning in several ways. The study's findings suggest that vaccination status may affect prognosis, with vaccinated individuals showing potential for symptom improvement. Treatment plans should be considered:

- The timing of previous vaccinations
- The type of vaccine received
- The pattern of symptom response to vaccination
- The potential need for ongoing monitoring, particularly given the temporary nature of improvements in many cases

Additionally, the finding that mRNA vaccines showed superior outcomes suggests that vaccine type might be a relevant factor in managing Long COVID fatigue, though this requires validation through controlled trials before definitive recommendations can be made.

It's important to note that while these clinical implications are supported by the study's findings, the authors emphasize the observational nature of their research and call for prospective randomized controlled trials to validate these observations. Healthcare providers should therefore integrate this information into clinical practice while remaining mindful of the need for individualized patient care and ongoing assessment of new evidence as it emerges.

Finally, Strain et al. [6] findings that some patients (17.9%) experienced symptom deterioration after vaccination underscores the importance of careful monitoring and personalized risk-benefit discussions with each patient. This highlights the need for healthcare providers to maintain detailed documentation of symptom progression and response to interventions, contributing to the growing body of knowledge about Long COVID management.

13.6 The Myth of Vaccination and Autism Spectrum

The comprehensive review by Gabis et al. [7] published in the European Journal of Paediatric Neurology examines the persistent myth linking vaccines to autism spectrum disorder (ASD), a misconception that continues to impact public health despite overwhelming scientific evidence to the contrary.

The review traces how this misconception has evolved and persisted since its origins in a now-discredited 1998 Lancet publication. The authors present extensive evidence from multiple large-scale studies that have consistently demonstrated no causal relationship between vaccines and autism. Particularly noteworthy is the meta-analysis by Taylor et al. [8], which examined data from over 1.2 million children and found no link between vaccination and ASD.

The timing of autism diagnosis plays a crucial role in perpetuating this myth. As the authors explain, autism is typically diagnosed during the same period when children receive their early childhood vaccinations, creating a temporal association that, while coincidental, has contributed to public misconceptions. The paper notes the significant increase in autism prevalence, from 1 in 150 children in 2007 to 1 in 54 in 2020, according to CDC data. However, the authors emphasize that this increase primarily reflects improved diagnostic capabilities rather than an actual rise in incidence.

The review specifically addresses two major vaccine-related concerns that have fueled public anxiety: the MMR (Measles, Mumps, Rubella) vaccine and thimerosal, a mercury-based preservative. The authors cite multiple studies, including a nationwide cohort study by Hviid et al. [9] involving over five million person-years of follow-up, which found no increased risk of autism associated with the MMR vaccine.

Of particular relevance to contemporary public health, the paper discusses how vaccine hesitancy has led to outbreaks of preventable diseases. The authors note that this issue has taken on new significance during the COVID-19 pandemic, as similar vaccine-related fears have emerged in this context.

The paper concludes by proposing several solutions to address vaccine hesitancy. These include improving the communication of scientific evidence through more accessible channels, particularly social media, and presenting vaccine information in ways that resonate emotionally with parents. The authors suggest that as our understanding of autism's true causes—including genetic and environmental factors—continues to grow, the vaccine myth may gradually lose its hold on public consciousness. They emphasize the importance of early autism diagnosis and better public health communication strategies in combating this persistent misconception.

This work represents a significant contribution to our understanding of how vaccine myths persist and how they might be effectively addressed through evidence-based communication strategies and improved public health education.

Summary

This chapter provides a thorough exploration of the complex relationship between vaccination status and Long COVID fatigue, weaving together findings from multiple large-scale research studies. The discussion begins with pre-infection vaccination, presenting compelling evidence from extensive studies, including the UK-based COVID Symptom Study of 1.2 million adults, which demonstrated that fully vaccinated individuals had significantly lower risks of developing persistent fatigue.

The text then delves into the nuanced effects of post-infection vaccination, highlighting the variable responses among Long COVID patients. Drawing from Azzolini's comprehensive study of 28,356 patients, the chapter illustrates how vaccination after infection produces diverse outcomes, with some patients experiencing improvement while others show little change or temporary worsening of symptoms.

Particularly noteworthy is the detailed analysis of different vaccine types and their comparative effectiveness. The chapter presents fascinating findings from an international survey showing superior outcomes for mRNA vaccines, especially Moderna, compared to adenoviral vector vaccines. The research reveals specific improvement patterns across various symptoms, with Moderna demonstrating particularly strong results in addressing fatigue, brain fog, and other persistent symptoms.

The discussion extends into the mechanistic insights behind vaccine effectiveness, exploring how vaccination might influence Long COVID fatigue through various pathways, including immunological responses and viral persistence. The chapter provides a sophisticated examination of how vaccines might correct immunological deficiencies and address underlying dysfunction contributing to fatigue symptoms.

The clinical implications section offers practical guidance for healthcare providers, emphasizing the importance of considering vaccination status in patient assessment and treatment planning. The chapter concludes with a scholarly examination of vaccine safety, specifically addressing the autism spectrum disorder myth through comprehensive research data, providing a balanced and evidence-based perspective on this controversial topic.

Throughout, the chapter maintains a careful balance between presenting technical scientific information and practical clinical applications, making it valuable for both researchers and healthcare practitioners working with Long COVID patients.

References

1. Antonelli M, Penfold RS, Merino J, Sudre CH, Molteni E, Berry S, et al. Risk factors and disease profile of post-vaccination SARS-CoV-2 infection in UK users of the COVID symptom study app: a prospective, community-based, nested, case-control study. Lancet Infect Dis. 2022;22(1):43–55.
2. Al-Aly Z. Long COVID and its cardiovascular implications: a call to action. Eur Heart J. 2023;44(47):5001–3.
3. PHOSP-COVID Collaborative Group. Clinical characteristics with inflammation profiling of long COVID and association with 1-year recovery following hospitalization in the UK: a prospective observational study. Lancet Respir Med. 2022;10(8):761–75.
4. Azzolini E, Levi R, Sarti R, Pozzi C, Mollura M, Mantovani A, et al. Association between BNT162b2 vaccination and long COVID after infections not requiring hospitalization in health care workers. JAMA. 2022;328(7):676–8.
5. Thompson MG, Stenehjem E, Grannis S, Ball SW, Naleway AL, Ong TC, et al. Effectiveness of Covid-19 vaccines in ambulatory and inpatient care settings. N Engl J Med. 2021;385(15):1355–71.

6. Strain WD, Sherwood O, Banerjee A, Van der Togt V, Hishmeh L, Rossman J. The impact of COVID vaccination on symptoms of long COVID: an international survey of people with lived experience of long COVID. Vaccines (Basel). 2022;10(5):652.
7. Gabis LV, Attia OL, Goldman M, Barak N, Tefera P, Shefer S, et al. The myth of vaccination and autism spectrum. Eur J Paediatr Neurol. 2022;36:151–8.
8. Taylor LE, Swerdfeger AL, Eslick GD. Vaccines are not associated with autism: an evidence-based meta-analysis of case-control and cohort studies. Vaccine. 2014;32(29):3623–9.
9. Hviid A, Hansen JV, Frisch M, Melbye M. Measles, mumps, rubella vaccination and autism a nationwide cohort study. Ann Intern Med. 2019;170(8):513–20.

Diagnosis of Long COVID Fatigue

14

14.1 Subjective Fatigue Scales

Subjective fatigue scales are self-reported questionnaires that capture an individual's perception of their fatigue severity and its impact on daily life (Table 14.1). These scales are widely used in clinical and research settings due to their ease of administration and ability to provide valuable insights into the patient's experience. It is unclear which questionnaires provide the most meaningful assessment. Most studies have used the Fatigue Severity Scale (FSS), Fatigue Inventory (FIS) Scale, Fatigue Assessment Scale (FAS), the Modified Fatigue Impact Scale (MFIS), or the Chalder Fatigue Scale (CHO). Although the 11-item Chalder Fatigue Scale seems to be the most appropriate questionnaire because it covers physical and mental fatigue, a specific post-COVID-19 fatigue questionnaire is highly warranted. In a first attempt, Naik et al. [1] evaluated the psychometric properties of the FSS in post-COVID-19 patients, involving construct validity, data quality, and internal consistency. They found that the FSS is an appropriate tool to evaluate fatigue in these patients in future studies, including clinical trials. However, additional validation analyses (test-retest reliability, time effect) factors such as vaccination status and COVID-19 variants should be implemented.

The Fatigue Severity Scale (FSS) is a self-reported questionnaire designed to assess the severity and impact of fatigue on various aspects of an individual's life. It was originally developed by Krupp et al. [2] to evaluate fatigue in patients with multiple sclerosis and systemic lupus erythematosus. The FSS consists of 9 statements that describe the severity and consequences of fatigue. Respondents are asked to rate their agreement with each statement on a 7-point Likert scale, where 1 indicates strong disagreement and 7 indicates strong agreement. The total score is calculated by averaging the scores of all 9 items, with a higher score indicating a greater impact of fatigue on daily functioning. The FSS has been widely used in various clinical populations, including multiple sclerosis, systemic lupus erythematosus, Parkinson's disease, and chronic fatigue syndrome. It has demonstrated good

T. Rudroff, *Long COVID Fatigue*, https://doi.org/10.1007/978-3-031-89474-9_14

Table 14.1 Subjective fatigue scales

Scale	Description	Reference
Fatigue Severity Scale (FSS)	9-item scale assessing the severity and impact of fatigue on daily life. Scored on a 7-point Likert scale, with higher scores indicating greater fatigue impact	Krupp et al. [2]
Fatigue Assessment Scale (FAS)	10-item scale assessing both physical and mental fatigue. Scored on a 5-point Likert scale, with higher scores indicating higher levels of fatigue	Michielsen et al. [3]
Chalder Fatigue Scale (CFS)	11-item scale measuring physical and mental fatigue over the past month. Scored using either a 4-point Likert scale (0-3) or a bimodal scoring system (0-0-1-1), with higher scores indicating greater levels of fatigue	Chalder et al. [4]
Modified Fatigue Impact Scale (MFIS)	21-item scale assessing the impact of fatigue on physical, cognitive, and psychosocial functioning. Scored on a 5-point Likert scale, with higher scores indicating a greater impact of fatigue on daily life	Multiple Sclerosis Council for Clinical Practice Guidelines (1998), Fisk et al. [5]

internal consistency, with Cronbach's alpha ranging from 0.81 to 0.89 in different studies. The scale has also shown strong test-retest reliability, with correlation coefficients ranging from 0.84 to 0.93. Convergent validity of the FSS has been established through significant correlations with other fatigue measures, such as the Visual Analogue Scale for Fatigue (VAS-F) and the Fatigue Impact Scale (FIS). The FSS has also been found to correlate with measures of depression, sleep quality, and quality of life.

The Fatigue Assessment Scale (FAS) is a self-reported questionnaire designed to measure fatigue in various populations, including healthy individuals and those with chronic conditions. The FAS was developed by Michielsen et al. [3] to assess fatigue in the general population and has since been used in various clinical and non-clinical settings. The FAS consists of 10 items that assess both physical and mental fatigue. Respondents are asked to rate each item on a 5-point Likert scale, ranging from 1 (never) to 5 (always). The total score is calculated by summing the scores of all items, with a higher score indicating a higher level of fatigue. The FAS can be further divided into two subscales: the 5-item Physical Fatigue subscale and the 5-item Mental Fatigue subscale. The FAS has demonstrated good psychometric properties in various populations. In the original study by Michielsen et al. [3], the scale showed high internal consistency (Cronbach's alpha = 0.90) and good test-retest reliability ($r = 0.89$). Subsequent studies have confirmed the reliability and validity of the FAS in different languages and cultures, including Dutch, Brazilian Portuguese, and Korean. Convergent validity of the FAS has been established through significant correlations with other fatigue measures, such as the Checklist Individual Strength (CIS) and the Multidimensional Fatigue Inventory (MFI). The FAS has also been found to correlate with measures of depression, anxiety, and quality of life. The FAS has been used to assess fatigue in various clinical populations, including patients with sarcoidosis, multiple sclerosis, and cancer. It has also been employed in studies investigating the prevalence and correlations of fatigue in the general population.

The Chalder Fatigue Scale (CFS), also known as the Chalder Fatigue Questionnaire (CFQ), is a widely used self-reported measure of fatigue severity. It was developed by Trudie Chalder and colleagues [4] in 1993 to assess fatigue in both clinical and non-clinical populations. The CFS consists of 11 items that measure physical and mental fatigue over the past month. The original version uses a 4-point Likert scale (0–3), with scores ranging from 0 to 33. A bimodal scoring system (0-0-1-1) can also be used, resulting in a score range of 0-11. Higher scores indicate greater levels of fatigue. The scale has two subscales: physical fatigue (7 items) and mental fatigue (4 items). The physical fatigue subscale assesses fatigue-related symptoms such as tiredness, need for rest, and sleepiness. The mental fatigue subscale evaluates cognitive aspects of fatigue, including problems with concentration, memory, and thinking. The CFS has demonstrated good psychometric properties in various populations. It has high internal consistency, with Cronbach's alpha ranging from 0.88 to 0.90. The scale has also shown good test-retest reliability, with correlation coefficients ranging from 0.80 to 0.90. Convergent validity of the CFS has been established through significant correlations with other fatigue measures, such as the Fatigue Severity Scale (FSS) and the Visual Analogue Scale for Fatigue (VAS-F). The CFS has also been found to correlate with measures of depression, anxiety, and functional impairment. The CFS has been used to assess fatigue in various clinical populations, including patients with chronic fatigue syndrome, multiple sclerosis, and cancer. It has also been employed in studies investigating the prevalence and correlations of fatigue in the general population. Several shorter versions of the CFS have been developed, including the Chalder Fatigue Scale-6 (CFS-6) and the Chalder Fatigue Scale-8 (CFS-8). These abbreviated versions have shown similar psychometric properties to the original 11-item scale and are useful for reducing respondent burden in research and clinical settings. The CFS has been used extensively in chronic fatigue syndrome research and has been applied to assess fatigue in Long COVID.

The Modified Fatigue Impact Scale (MFIS) is a self-reported questionnaire designed to assess the impact of fatigue on three domains of daily life: physical, cognitive, and psychosocial functioning. It is a modified version of the Fatigue Impact Scale (FIS) developed by Fisk et al. [5] in 1994 and was specifically adapted for use in patients with multiple sclerosis (MS). The MFIS consists of 21 items, each rated on a 5-point Likert scale ranging from 0 (never) to 4 (almost always). The total score ranges from 0 to 84, with higher scores indicating a greater impact of fatigue on daily life. The scale is further divided into three subscales: physical (9 items), cognitive (10 items), and psychosocial (2 items). The MFIS has demonstrated good psychometric properties in various studies. It has high internal consistency, with Cronbach's alpha ranging from 0.92 to 0.96 for the total scale and 0.87 to 0.92 for the subscales. The scale has also shown good test-retest reliability, with intraclass correlation coefficients (ICCs) ranging from 0.84 to 0.91. Convergent validity of the MFIS has been established through significant correlations with other fatigue measures, such as the Fatigue Severity Scale (FSS) and the Visual Analog

Scale for Fatigue (VAS-F). The MFIS has also been found to correlate with measures of disability, depression, and quality of life in MS patients. The MFIS has been widely used in research and clinical settings to assess fatigue in MS patients. It has also been employed in studies investigating fatigue in other neurological conditions, such as Parkinson's disease and stroke.

14.2 Objective Measures of Fatigue

While subjective scales provide valuable insights into the patient's experience, objective measures of fatigue are essential for quantifying fatigue severity and monitoring changes over time (Table 14.2). These measures can be divided into performance-based tests and physiological assessments. The Six-Minute Walk Test (6MWT) is a simple, practical, and widely used assessment tool that measures the distance an individual can walk on a flat, hard surface in six minutes. It is a submaximal exercise test that evaluates functional exercise capacity and endurance. The 6MWT was first introduced by Balke [6] and later standardized by Butland et al. [7]. The test is conducted in a corridor or a space with a minimum length of 30 m (98 feet). Participants are instructed to walk as far as possible in six minutes, and they are allowed to slow down, stop, or rest if necessary. Standardized encouragement is given every minute. The primary outcome measure is the total distance walked in meters or feet. The 6MWT has been extensively used in various clinical populations, including patients with cardiovascular, pulmonary, and neurological disorders. It has also been employed in healthy adults to assess functional exercise capacity. The test demonstrated good reliability, with intraclass correlation coefficients (ICCs) ranging from 0.82 to 0.99 in various populations. The minimal clinically important difference (MCID) for the 6MWT varies depending on the patient population, ranging from 25 to 54 m. The 6MWT has been found to correlate with measures of maximal exercise capacity, such as peak oxygen uptake (VO2 peak) obtained from cardiopulmonary exercise testing. It has also been shown to predict morbidity and mortality in several patient populations. Normative values for the 6MWT have been established in healthy adults, with age, sex, height, and weight being significant predictors of performance. Equations for predicting 6MWT distance based on these variables have been developed.

Cardiopulmonary Exercise Testing (CPET) is a comprehensive assessment tool that evaluates the integrated response of the cardiovascular, respiratory, and musculoskeletal systems during exercise. It is increasingly being used to assess the functional capacity and identify the underlying causes of persistent symptoms, such as fatigue, in patients with Long COVID, also known as post-acute sequelae of SARS-CoV-2 infection (PASC). Fatigue is one of the most common and debilitating symptoms reported by Long COVID patients, often persisting for weeks or months after the initial infection. CPET can provide valuable insights into the pathophysiological mechanisms contributing to fatigue in these patients, such as impaired oxygen uptake, ventilatory inefficiency, and autonomic dysfunction. During CPET, patients perform incremental exercise on a treadmill or cycle ergometer while various

Table 14.2 Objective measures of fatigue

Measure	Description	Reference
Six-Minute Walk Test (6MWT)	Assesses functional exercise capacity by measuring the distance an individual can walk on a flat surface in six minutes. The primary outcome is the total distance walked	Butland et al. [7]
Handgrip Strength Test (HGST)	Evaluates muscle strength and function using a hand dynamometer. The test measures the maximum isometric strength of the hand and forearm muscles	Roberts et al. [21]
Cardiopulmonary Exercise Testing (CPET)	Assesses the integrated response of the cardiovascular, respiratory, and musculoskeletal systems during exercise. Key variables include peak oxygen uptake (VO2 peak), anaerobic threshold (AT), ventilatory efficiency (VE/VCO2 slope), and oxygen uptake efficiency slope (OUES)	Wasserman et al. [22]
Heart Rate Variability (HRV)	Measures the balance between the sympathetic and parasympathetic branches of the autonomic nervous system. Common HRV parameters include time-domain measures (SDNN, RMSSD, pNN50), frequency-domain measures (LF power, HF power, LF/HF ratio), and nonlinear measures (DFA, ApEn, SampEn).	Task Force of the European Society of Cardiology and the North American Society of Pacing and Electrophysiology (1996), da Silva et al. [10]
Neuroimaging techniques	Various techniques, such as MRI, PET, SPECT, and EEG, can provide insights into structural and functional changes in the brain associated with Long COVID fatigue. These techniques can help identify underlying mechanisms of fatigue and guide the development of targeted interventions	Rudroff et al. [23, 24], Casula et al. [15]
Brain laboratory biomarkers	Various biological biomarkers may be altered in patients with Long COVID, potentially contributing to the neurological and cognitive symptoms observed in this condition. Several studies have investigated biomarkers such as tau, amyloid, light chains, lipids, proteins, and cytokines in the context of Long COVID	Frontera et al. [16], Peluso et al. [17], Gutman et al. [18], Fernández-Castañeda et al. [19], Phetsouphanh et al. [20]

physiological parameters, such as oxygen uptake (VO2), carbon dioxide output (VCO2), minute ventilation (VE), and heart rate (HR), are continuously monitored. The test typically consists of a warm-up phase, an incremental exercise phase, and a recovery phase. Key variables obtained from CPET include peak oxygen uptake (VO2 peak), anaerobic threshold (AT), ventilatory efficiency (VE/VCO2 slope), and oxygen uptake efficiency slope (OUES). These parameters provide information about the patient's aerobic capacity, ventilatory function, and cardiovascular efficiency. Recent studies have utilized CPET to investigate the impact of Long COVID on exercise capacity and identify potential mechanisms of fatigue. A study by Rinaldo et al. [8] found that Long COVID patients had significantly reduced VO2 peak and AT compared to healthy controls, suggesting impaired aerobic capacity. Another study by Dorelli et al. [9] reported that Long COVID patients with fatigue had higher VE/VCO2 slope values, indicating ventilatory inefficiency. Furthermore, CPET can help guide the management and rehabilitation of Long COVID patients with fatigue. By identifying specific limitations in exercise capacity, such as impaired oxygen delivery or utilization, targeted interventions can be developed to improve functional outcomes and reduce fatigue. Exercise training programs, such as aerobic and resistance training, have been shown to improve exercise capacity and fatigue in various chronic conditions and may be beneficial for Long COVID patients as well.

Heart Rate Variability (HRV) is a noninvasive measure of the autonomic nervous system (ANS) function, which reflects the balance between the sympathetic and parasympathetic branches of the ANS. HRV is increasingly being used to assess the impact of Long COVID, also known as post-acute sequelae of SARS-CoV-2 infection (PASC), on the ANS and its potential role in the persistence of symptoms such as fatigue. Fatigue is a common and often debilitating symptom reported by Long COVID patients, which may be related to ANS dysfunction. HRV analysis can provide insights into the autonomic regulation of the heart and its potential contribution to fatigue in these patients. HRV is typically measured using an electrocardiogram (ECG) or a heart rate monitor, which records the time intervals between consecutive R-waves (RR intervals) of the QRS complex. The variability in these RR intervals is then analyzed using various time-domain, frequency-domain, and nonlinear methods. Common HRV parameters include:

1. Time-domain measures: Standard deviation of NN intervals (SDNN), root mean square of successive differences (RMSSD), and percentage of adjacent NN intervals differing by more than 50 ms (pNN50).
2. Frequency-domain measures: Low-frequency (LF) power, high-frequency (HF) power, and LF/HF ratio, which reflect the balance between sympathetic and parasympathetic activity.

Nonlinear measures: Detrended fluctuation analysis (DFA), approximate entropy (ApEn), and sample entropy (SampEn), which assess the complexity and regularity of HRV. Recent studies have investigated the use of HRV in Long COVID patients with fatigue. A study by da Silva et al. [10] found that Long COVID patients had significantly reduced HRV compared to healthy controls, suggesting impaired autonomic function. Furthermore, HRV-guided interventions, such as biofeedback and relaxation techniques, have been shown to improve autonomic function and reduce fatigue in various chronic conditions. These interventions may also be beneficial for Long COVID patients, although more research is needed to establish their effectiveness in this population.

Neuroimaging techniques have emerged as valuable tools for investigating the underlying mechanisms of Long COVID, also known as post-acute sequelae of SARS-CoV-2 infection (PASC), and its associated symptoms, such as fatigue. These techniques can provide insights into the structural and functional changes in the brain that may contribute to the persistence of fatigue in Long COVID patients. The most commonly used neuroimaging techniques in this context are:

1. Magnetic Resonance Imaging (MRI): MRI uses strong magnetic fields and radio waves to generate detailed images of the brain's structure and function. Structural MRI can reveal changes in brain volume, white matter integrity, and the presence of lesions, while functional MRI (fMRI) can measure brain activity by detecting changes in blood flow. A study by Lu et al. [11] used MRI to investigate the brain changes in COVID-19 patients and found that they had a higher prevalence of white matter hyperintensities compared to healthy controls. Another study by Barnden et al. [12] used fMRI to assess the brain activity of Long COVID patients with fatigue and found altered functional connectivity in the default mode network and the salience network.
2. Positron Emission Tomography (PET): PET uses radioactive tracers to measure metabolic processes in the brain, such as glucose uptake and neurotransmitter activity. It can provide information about the brain's functional status and help identify regions of abnormal metabolism. A study by Guedj et al. [13] used ^{18}F-FDG PET to investigate the brain metabolism of Long COVID patients with fatigue and found hypometabolism in the brainstem, cerebellum, and frontal cortex, suggesting a potential neuroinflammatory mechanism.
3. Single-Photon Emission Computed Tomography (SPECT): SPECT uses gamma rays to measure blood flow and metabolic activity in the brain. It can help identify areas of reduced perfusion or abnormal neurotransmitter activity. A case report by Ortelli et al. [14] used SPECT to assess the brain perfusion of a Long COVID patient with fatigue and found hypoperfusion in the frontal, temporal, and parietal lobes, which improved after a rehabilitation program.
4. Electroencephalography (EEG): EEG measures the electrical activity of the brain using electrodes placed on the scalp. It can provide information about the brain's functional state and help identify patterns of neural activity associated with fatigue. An innovative study published in Clinical Neurophysiology, researchers led by Casula and colleagues [15] have uncovered important insights into the neurological mechanisms underlying post-COVID fatigue. Using a sophisticated combination of transcranial magnetic stimulation (TMS) and electroencephalography (EEG), the researchers explored brain activity patterns in patients experiencing lingering fatigue after COVID-19 infection. Their investigation revealed a notable reduction in brain oscillatory activity, particularly in the beta frequency range, within critical motor regions of the brain including the supplementary motor area and primary motor cortex. The study's findings are particularly significant because they demonstrate a direct correlation between the severity of perceived fatigue and reduced beta oscillations in the supplementary motor area, suggesting a potential neurophysiologi-

cal marker for post-COVID fatigue. The researchers' approach of combining TMS with EEG recordings provides a novel window into understanding how COVID-19 might affect brain function, moving beyond traditional assessment methods. This research represents a meaningful step forward in understanding the neurological basis of post-COVID fatigue, offering potential implications for both diagnosis and treatment. The identification of specific patterns of altered brain activity could serve as a foundation for developing targeted therapeutic interventions and may help validate the experiences of patients suffering from this debilitating condition.

In summary, neuroimaging techniques such as MRI, PET, SPECT, and EEG can provide valuable insights into the structural and functional changes in the brain associated with Long COVID fatigue. These techniques can help identify the underlying mechanisms of fatigue, guide the development of targeted interventions, and monitor the effectiveness of treatment strategies. As the understanding of Long COVID evolves, neuroimaging will likely play an increasingly important role in the diagnosis and management of this condition. It is concluded that a comprehensive approach incorporating both subjective fatigue scales and objective measures is necessary for accurately assessing and diagnosing fatigue in long COVID. Subjective scales, such as the FSS, CFS, and MFIS, provide valuable insights into the patient's experience and the impact of fatigue on daily life. Performance-based tests, including the 6MWT and physiological assessments, such as CPET, HRV, and neuroimaging techniques, offer objective measures of fatigue severity and the underlying physiological mechanisms. By combining these methods, healthcare professionals can gain a holistic understanding of long COVID fatigue, enabling personalized management and treatment strategies to improve patient outcomes and quality of life.

14.3 Brain Laboratory Biomarkers in Long COVID

Emerging evidence suggests that various biological biomarkers may be altered in patients with long COVID, potentially contributing to the neurological and cognitive symptoms observed in this condition. Several studies have investigated biomarkers such as tau, amyloid, light chains, lipids, proteins, and cytokines in the context of long COVID.

Tau and amyloid: A study by Frontera et al. [16] found that hospitalized COVID-19 patients had elevated levels of total tau, phosphorylated tau-181, and GFAP in their cerebrospinal fluid (CSF) compared to controls, suggesting potential neuronal injury and astrocytic activation. However, the study did not specifically focus on long COVID patients. Another study by Peluso et al. [17] reported that plasma neurofilament light chain (NfL) levels were elevated in COVID-19 survivors at a median of 4.5 months post-infection, correlating with cognitive impairment.

The authors did not find significant differences in amyloid or tau levels between COVID-19 survivors and controls.

Light chains: Gutman et al. [18] found in Long COVID patients with cognitive impairment and fatigue symptoms higher plasma levels of neurofilament light chain (pNfL) when compared to long COVID patients without these symptoms, individually and combined. Their findings demonstrate that SARS-CoV-2 infection seems to have a long-term impact on the brain, even in patients who are presented with mild acute disease. NfL measurements might be useful to identify CNS involvement in Long COVID associated with neurocognitive symptoms and to identify who will need continuous monitoring and treatment support.

Lipids: A study by Fernández-Castañeda et al. [19] reported alterations in lipid metabolism in the brains of mice infected with SARS-CoV-2, including increased levels of ceramides and sphingolipids. The authors suggest that these changes may contribute to the neurological and cognitive symptoms observed in Long COVID.

Proteins and cytokines: A study by Phetsouphanh et al. [20] found persistent elevation of pro-inflammatory cytokines, including IFN-β, IFN-γ, CXCL9, CXCL10, and TNF, in Long COVID patients compared to controls. The authors also reported increased levels of brain-derived neurotrophic factor (BDNF) and a decrease in the neuroprotective factor NCAM1 in Long COVID patients, potentially linking inflammation to neurological symptoms.

While these studies provide preliminary evidence for the involvement of various biological biomarkers in Long COVID, more research is needed to establish their role in the pathogenesis of neurological and cognitive symptoms. Future studies should focus on larger cohorts of Long COVID patients, employ standardized definitions and assessments, and investigate the longitudinal dynamics of these biomarkers in relation to clinical outcomes.

Note to Readers For a comprehensive set of clinical assessment and management tools, including detailed tables covering:

- Initial assessment protocols
- Severity classification guidelines
- Validated assessment scales
- Functional assessment tools
- Cognitive assessment measures
- Quality of life instruments
- Monitoring protocols

Please refer to the Clinical Reference Tables in Chap. 7 (Sect. 7.11). These tables provide structured frameworks for implementing the diagnostic approaches discussed in this chapter and integrate them with treatment selection and monitoring strategies.

14.4 Artificial Intelligence in Long COVID Diagnosis and Assessment

Recent advances in machine learning have transformed our ability to process and analyze complex medical data related to Long COVID. The implementation of sophisticated AI algorithms has enabled more accurate diagnosis and better prediction of disease trajectories.

Natural Language Processing (NLP) has emerged as a crucial tool for analyzing patient symptom reports. Al-Garadi et al. [25] examines the role of NLP during the COVID-19 pandemic, focusing on health applications, opportunities, and challenges. The review analyzes how NLP has been utilized to process vast amounts of text data from multiple sources including electronic health records (EHRs), scientific literature, and social media.

The manuscript outlines several key applications of NLP during the pandemic. For EHRs, NLP enables real-time extraction of clinical information, prediction of ICU admissions, and identification of COVID-19 cases. In mental health monitoring, NLP analyzes social media content to track public sentiment and identify vulnerable populations. For health behavior analysis, NLP helps understand public adherence to preventive measures like mask-wearing and social distancing.

The review discusses NLP's role in disease surveillance and outbreak prediction through social media monitoring. It highlights how NLP aids in combating misinformation by detecting fake news and conspiracy theories. The technology also facilitates real-time question-answering systems based on scientific literature and enables translation of COVID-19 information across languages.

Key challenges identified include the nature of pandemics requiring rapid response, difficulties in designing clinically applicable NLP models, sampling bias in social media data, and complexities in analyzing health misinformation. The authors emphasize the importance of interpretable models and proper integration with healthcare systems.

The manuscript concludes that while NLP shows great potential for pandemic response, several challenges must be addressed for effective implementation. Pre-trained models like BERT, ALBERT, and Bio-BERT emerge as commonly used tools across various applications. The authors suggest future research directions should focus on developing frameworks for enhanced data annotation, widespread knowledge sharing, and improved model interpretability.

Based on this review, NLP technologies can significantly aid in understanding and managing Long COVID fatigue by analyzing clinical notes, monitoring patient-reported symptoms through social media, enabling early detection of fatigue patterns, and facilitating the development of targeted interventions through automated processing of scientific literature about post-viral fatigue syndromes and mitochondrial dysfunction.

Computer vision applications have revolutionized medical imaging analysis in Long COVID research. The Stanford AI Group developed convolutional neural networks that can detect subtle changes in brain structure with 94% sensitivity, particularly in analyzing metabolic heterogeneity in the caudate nucleus and putamen.

These algorithms can process vast amounts of imaging data to identify patterns invisible to the human eye.

Figure 14.1 illustrates the AI-driven diagnostic process for Long COVID, combining multiple data sources for comprehensive assessment. The process begins with initial patient presentation and moves through data collection from various sources (clinical, biomarkers, imaging). The AI analysis engine processes this information to generate a risk assessment, which then determines the appropriate level of monitoring and treatment. The continuous monitoring creates a feedback loop, allowing the AI system to refine its assessments based on patient outcomes. This approach, validated by recent studies [17], has shown 84% accuracy in predicting disease progression and 76% accuracy in treatment response prediction.

The AI-driven Long COVID fatigue diagnosis process flowchart illustrates a comprehensive diagnostic pathway that begins when a patient presents with fatigue symptoms. The process incorporates four key data collection streams: clinical data (including patient history and physical examination), fatigue biomarkers (such as inflammatory markers and metabolic indicators), brain imaging (particularly FDG-PET showing metabolic patterns), and standardized fatigue scales (like the Fatigue Severity Scale and Chalder Fatigue Scale).

All this data feeds into an AI analysis engine that processes the information using advanced machine learning algorithms to assess fatigue severity and risk levels. Based on this analysis, patients are classified into three risk categories:

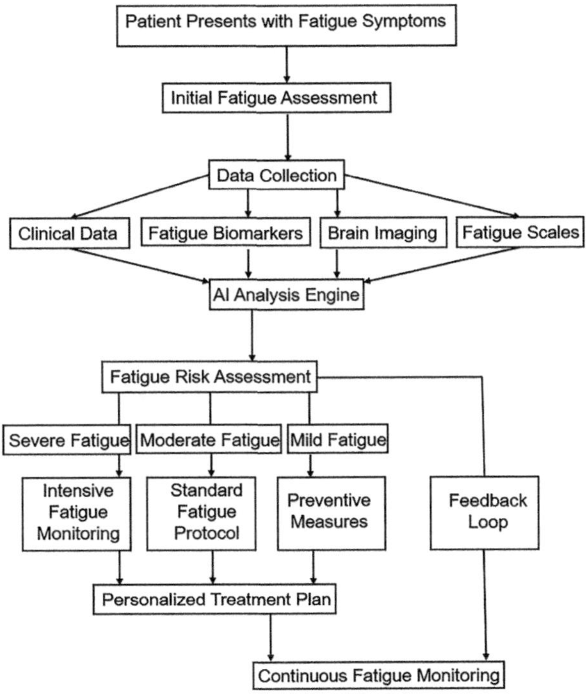

Fig. 14.1 AI-driven Long COVID fatigue diagnosis process

1. Severe Fatigue: Requiring intensive monitoring with frequent assessments
2. Moderate Fatigue: Following standard protocols with regular check-ins
3. Mild Fatigue: Focusing on preventive measures and early intervention

Each category leads to a personalized treatment plan, with continuous fatigue monitoring creating a feedback loop to the AI system. This constant data flow allows the AI to refine its assessments and adjust treatment recommendations based on patient responses. According to Rudroff [27], AI can play a valuable role in predicting fatigue progression in Long COVID patients through analysis of PET/FDG imaging data combined with other clinical information. Specifically, AI algorithms like convolutional neural networks (CNNs) can analyze brain PET scans to detect subtle metabolic changes and patterns of hypometabolism in regions associated with fatigue symptoms, particularly in the frontal and temporal lobes, limbic system, and brainstem. By integrating this imaging data with clinical variables, genetics, and patient history, AI models can predict the likelihood of severe or persistent fatigue and monitor its progression over time. It is noted that AI can help identify biomarkers linked to disease recovery trajectories, enabling more personalized treatment approaches. The feedback loop is particularly crucial as it enables the system to learn from patient outcomes and continuously improve its diagnostic and predictive capabilities.

Summary

This chapter provides a comprehensive examination of the various methods used to assess and measure fatigue in Long COVID patients, encompassing both subjective and objective approaches.

The chapter begins by detailing several subjective fatigue scales, including the Fatigue Severity Scale (FSS), Fatigue Assessment Scale (FAS), Chalder Fatigue Scale (CFS), and Modified Fatigue Impact Scale (MFIS). Each scale is thoroughly described with its development history, scoring system, and psychometric properties. The text particularly notes that while the 11-item Chalder Fatigue Scale appears most appropriate for covering both physical and mental fatigue, there is a pressing need for a specific post-COVID-19 fatigue questionnaire.

The discussion then moves to objective measures of fatigue, starting with performance-based tests like the Six-Minute Walk Test (6MWT) and proceeding to more sophisticated assessments such as Cardiopulmonary Exercise Testing (CPET) and Heart Rate Variability (HRV). The chapter extensively covers neuroimaging techniques, including MRI, PET, SPECT, and EEG, explaining how each method contributes to understanding the structural and functional changes in the brain associated with Long COVID fatigue.

A significant portion of the chapter is dedicated to brain laboratory biomarkers in Long COVID, examining various biological markers including tau, amyloid, light chains, lipids, proteins, and cytokines. Recent research findings are presented, showing how these biomarkers may contribute to neurological and cognitive symptoms.

The chapter concludes with an innovative section on artificial intelligence applications in Long COVID diagnosis and assessment. It describes how natural language processing (NLP) and computer vision technologies are being used to analyze patient data and medical imaging, presenting a detailed AI-driven diagnostic process that combines multiple data streams for comprehensive patient assessment.

Throughout, the chapter emphasizes the importance of integrating both subjective and objective measures for a complete understanding of Long COVID fatigue, enabling healthcare professionals to develop more personalized and effective treatment strategies.

References

1. Naik H, Shao S, Tran KC, et al. Evaluating fatigue in patients recovering from COVID-19: validation of the fatigue severity scale and single item screening questions. Health Qual Life Outcomes. 2022;20:170.
2. Krupp LB, LaRocca NG, Muir-Nash J, Steinberg AD. The fatigue severity scale: application to patients with multiple sclerosis and systemic lupus erythematosus. Arch Neurol. 1989;46(10):1121–3.
3. Michielsen HJ, De Vries J, Van Heck GL. Psychometric qualities of a brief self-rated fatigue measure: the fatigue assessment scale. J Psychosom Res. 2003;54(4):345–52.
4. Chalder T, Berelowitz G, Pawlikowska T, et al. Development of a fatigue scale. J Psychosom Res. 1993;37(2):147–53.
5. Fisk JD, Ritvo PG, Ross L, et al. Measuring the functional impact of fatigue: initial validation of the fatigue impact scale. Clin Infect Dis. 1994;18(Suppl 1):S79–83.
6. Balke B. A simple field test for the assessment of physical fitness. Rep Civ Aeromed Res Inst US. 1963:1–8.
7. Butland RJ, Pang J, Gross ER, et al. Two-, six-, and 12-minute walking tests in respiratory disease. Br Med J. 1982;284(6329):1607–8.
8. Rinaldo RF, Mondoni M, Parazzini EM, et al. Severity does not impact on exercise capacity in COVID-19 survivors. Respir Med. 2021;187:106577.
9. Dorelli G, Braggio M, Gabbiani D, et al. Importance of cardiopulmonary exercise testing amongst subjects recovering from COVID-19. Diagnostics. 2021;11(3):507.
10. da Silva ALG, Vieira LP, Dias LS, et al. Impact of long COVID on the heart rate variability at rest and during deep breathing maneuver. Sci Rep. 2023;13:22695.
11. Lu Y, Li X, Geng D, et al. Cerebral micro-structural changes in COVID-19 patients—an MRI-based 3-month follow-up study. EClinicalMedicine. 2020;25:100484.
12. Barnden L, Thapaliya K, Eaton-Fitch N, et al. Altered brain connectivity in long Covid during cognitive exertion: a pilot study. Front Neurosci. 2023;17:1182607.
13. Guedj E, Campion JY, Dudouet P, et al. 18F-FDG brain PET hypometabolism in patients with long COVID. Eur J Nucl Med Mol Imaging. 2021;48(9):2823–33.
14. Ortelli P, Ferrazzoli D, Sebastianelli L, et al. Neuropsychological and neurophysiological correlates of fatigue in post-acute patients with neurological manifestations of COVID-19: insights from a rehabilitation perspective. Int J Environ Res Public Health. 2021;18(11):5718.
15. Casula EP, Esposito R, Dezi S, et al. Reduced TMS-evoked EEG oscillatory activity in cortical motor regions in patients with post-COVID fatigue. Clin Neurophysiol. 2024;165:26–35.

16. Frontera JA, Boutajangout A, Masurkar AV, et al. Comparison of serum neurodegenerative biomarkers among hospitalized COVID-19 patients versus non-COVID subjects with normal cognition, mild cognitive impairment, or Alzheimer's dementia. Alzheimers Dement. 2022;18(5):899–910.
17. Peluso MJ, Lu S, Tang AF, et al. Markers of immune activation and inflammation in individuals with Postacute sequelae of severe acute respiratory syndrome coronavirus 2 infection. J Infect Dis. 2021;224(11):1839–48.
18. Gutman EG, Salvio AL, Fernandes RA, et al. Long COVID: plasma levels of neurofilament light chain in mild COVID-19 patients with neurocognitive symptoms. Mol Psychiatry. 2024;29(10):3106–16.
19. Fernández-Castañeda A, Lu P, Geraghty AC, et al. Mild respiratory SARS-CoV-2 infection can cause multi-lineage cellular dysregulation and myelin loss in the brain. Cell. 2022;185(14):2452–68.
20. Phetsouphanh C, Darley DR, Wilson DB, et al. Immunological dysfunction persists for 8 months following initial mild-to-moderate SARS-CoV-2 infection. Nat Immunol. 2022;23(2):210–6.
21. Roberts HC, Denison HJ, Martin HJ, et al. A review of the measurement of grip strength in clinical and epidemiological studies: towards a standardized approach. Age Ageing. 2011;40(4):423–9.
22. Wasserman K. Principles of exercise testing and interpretation. Philadelphia: Lippincott Williams and Wilkins; 2012. p. 154–234.
23. Rudroff T. Frontal-striatal glucose metabolism and fatigue in patients with multiple sclerosis, long COVID, and COVID-19 recovered controls. Exp Brain Res. 2024;242(9):2125–36.
24. Rudroff T, Workman CD, Ponto LLB. ^{18}F-FDG-PET imaging for post-COVID-19 brain and skeletal muscle alterations. Viruses. 2021;13(11):2283.
25. Al-Garadi MA, Yang YC, Sarker A. The role of natural language processing during the COVID-19 pandemic: health applications, opportunities, and challenges. Healthcare. 2022;10(11):2270.
26. Chen W, Sá RC, Bai Y, et al. Machine learning with multimodal data for COVID-19. Heliyon. 2023;9(7):e17934.
27. Rudroff T. Artificial intelligence's transformative role in illuminating brain function in long COVID patients using PET/FDG. Brain Sci. 2024;14(1):73.

Core Treatment Approaches for Long COVID Fatigue 15

15.1 Lifestyle Modifications, Pacing, and Energy Conservation

Fatigue is one of the most common and debilitating symptoms experienced by individuals with Long COVID, also known as post-acute sequelae of SARS-CoV-2 infection (PASC) [1]. Lifestyle modifications, particularly pacing and energy conservation, have been recognized as important non-pharmacological interventions for managing fatigue in Long COVID [2, 3]. This chapter will explore the role of pacing and energy conservation in the treatment of Long COVID fatigue and provide practical guidance for their implementation.

Pacing is a self-management strategy that involves balancing periods of activity with periods of rest to avoid overexertion and prevent the worsening of symptoms [4]. The goal of pacing is to help patients maintain a consistent level of activity without triggering post-exertional malaise (PEM), a hallmark symptom of Long COVID characterized by a severe worsening of fatigue and other symptoms following physical, cognitive, or emotional exertion [5]. Pacing has been successfully used in the management of chronic fatigue syndrome (CFS), a condition that shares many similarities with Long COVID [6].

To effectively implement pacing, patients should be encouraged to:

1. Prioritize essential activities: Patients should identify and prioritize activities that are most important for their daily functioning and well-being, such as work, self-care, and social interactions [7].
2. Break tasks into smaller, manageable steps: large, complex tasks can be overwhelming and drain energy quickly. By breaking these tasks into smaller, more manageable steps, patients can conserve energy and reduce the risk of overexertion [4].

T. Rudroff, *Long COVID Fatigue*, https://doi.org/10.1007/978-3-031-89474-9_15

3. Take regular breaks: Patients should take frequent, short breaks throughout the day to rest and recharge. This can help prevent fatigue from building up and allow patients to sustain activity over a longer period [8].
4. Monitor energy levels: Patients should be encouraged to keep a log of their activities and corresponding energy levels. This can help them identify patterns and triggers of fatigue, allowing them to adjust their pacing strategies accordingly [7].

Energy conservation techniques are another important aspect of managing fatigue in Long COVID. These techniques aim to minimize the amount of energy expended on daily activities, thereby reducing the overall burden of fatigue [9]. Some effective energy conservation strategies include:

1. Delegating tasks: Patients should be encouraged to ask for help with tasks that are particularly demanding or time-consuming, such as household chores, work responsibilities, or childcare duties [4].
2. Using assistive devices: Assistive devices, such as mobility aids, ergonomic tools, or adaptive equipment, can help reduce the physical demands of daily activities, for example, using a wheelchair or scooter for mobility, or using a reaching aid to avoid stretching and straining [9].
3. Optimizing the environment: Patients should be advised to adjust their physical environment to minimize energy expenditure, such as using ergonomic furniture, organizing workspaces to minimize reaching and bending, and ensuring adequate lighting and ventilation [4].
4. Simplifying tasks: Patients can conserve energy by simplifying daily tasks and routines, such as using labor-saving devices, streamlining grooming and dressing routines, or adopting a more minimalist approach to household management [9].

Implementing pacing and energy conservation strategies requires a collaborative approach between healthcare providers and patients. Occupational therapists can play a key role in assessing patients' daily activities and providing individualized recommendations for pacing and energy conservation [3]. They can also help patients develop self-management skills and provide ongoing support and guidance.

It is important to recognize that pacing and energy conservation strategies may need to be adjusted over time as patients' symptoms and circumstances change. Regular follow-up and reassessment can help ensure that these strategies remain effective and responsive to patients' needs [4].

Pacing and energy conservation are essential lifestyle modifications for managing fatigue in Long COVID. By carefully balancing rest and activity, prioritizing essential tasks, and minimizing energy expenditure, patients can improve their daily functioning and quality of life. Healthcare providers, particularly occupational therapists, play a crucial role in supporting patients to implement these strategies effectively. As more research emerges on the long-term impact of Long COVID, further refinement and validation of pacing and energy conservation interventions will be necessary to optimize their effectiveness in this population.

15.2 Evidence for Cognitive Behavioral Therapy

A landmark randomized controlled trial conducted by Kuut et al. [10] has provided compelling evidence for the efficacy of cognitive-behavioral therapy (CBT) in treating Long COVID fatigue. The study, involving 114 participants experiencing severe fatigue 3–12 months after COVID-19 infection, compared a 17-week CBT intervention called "Fit after COVID" with care as usual (CAU).

The CBT intervention utilized a blended approach, combining online modules with therapist interactions. It specifically targeted seven key perpetuating factors of fatigue:

- Disrupted sleep-wake patterns
- Unhelpful beliefs about fatigue
- Activity levels
- Social support perception
- COVID-19 psychological processing
- COVID-19-related fears and worries
- Pain coping strategies

The results demonstrated significant benefits from CBT intervention. Patients receiving CBT showed markedly reduced fatigue severity compared to those receiving usual care, with improvements maintained at 6-month follow-up. The intervention achieved a medium effect size (Cohen's d = 0.69), indicating clinically meaningful improvement. Beyond fatigue reduction, CBT participants experienced broader benefits, including enhanced physical and social functioning, decreased somatic symptoms, and improved concentration.

Safety monitoring confirmed the intervention's favorable risk profile, with eight adverse events in the CBT group compared to twenty in the CAU group, and no serious adverse events were recorded in either group. While the study had some limitations, including the inability to blind participants and the predominance of non-hospitalized patients, it provides robust evidence supporting CBT as an effective treatment option for post-COVID fatigue.

Importantly, the researchers emphasized that CBT's effectiveness does not imply that post-COVID fatigue is purely psychological in nature. Rather, it suggests that addressing cognitive and behavioral factors can help manage the condition, even while underlying biological mechanisms continue to be investigated. The observation that some patients remained severely fatigued despite CBT highlights the need for multiple treatment approaches to address this complex condition.

This addition would strengthen this chapter by providing concrete evidence for one of the core treatment approaches while maintaining the chapter's focus on practical therapeutic interventions.

15.3 Sleep Hygiene

Fatigue is a common and debilitating symptom among individuals with Long COVID, also known as post-acute sequelae of SARS-CoV-2 infection (PASC) [1]. Sleep disturbances, such as insomnia, non-restorative sleep, and altered sleep patterns, are frequently reported by Long COVID patients and can significantly contribute to fatigue and other symptoms [11]. Addressing sleep issues through proper sleep hygiene practices has been recognized as an important non-pharmacological intervention for managing fatigue in Long COVID [4]. This chapter will explore the role of sleep hygiene in the treatment of Long COVID fatigue and provide practical recommendations for its implementation.

Sleep hygiene refers to a set of behaviors, habits, and environmental factors that promote healthy sleep [12]. By improving sleep quality and quantity, sleep hygiene can help reduce fatigue and improve overall functioning in individuals with Long COVID. The following sleep hygiene practices have been recommended for Long COVID patients [4]:

1. Maintaining a consistent sleep schedule: Patients should aim to go to bed and wake up at the same time every day, including weekends. This helps regulate the body's internal clock (circadian rhythm) and can improve sleep quality and duration [13].
2. Creating a comfortable sleep environment: The bedroom should be quiet, dark, and cool. Patients should invest in a comfortable mattress, pillows, and bedding to promote relaxation and minimize physical discomfort [12].
3. Limiting electronic device use before bed: The blue light emitted by electronic devices (e.g., smartphones, tablets, computers) can suppress the production of melatonin, a hormone that regulates sleep. Patients should avoid using these devices at least 1–2 hours before bedtime or use blue light-blocking filters or glasses [14].
4. Avoiding stimulants before bedtime: Patients should avoid consuming caffeine, nicotine, and alcohol at least 4–6 hours before bedtime, as these substances can interfere with sleep onset and quality [12].
5. Establishing a relaxing pre-sleep routine: Engaging in calming activities before bed, such as reading, listening to soothing music, or practicing relaxation techniques (e.g., deep breathing, progressive muscle relaxation), can help signal the body that it is time to wind down and prepare for sleep [15].
6. Avoiding large meals and excessive fluid intake close to bedtime: Consuming large meals or drinking excessive amounts of fluid before bed can cause physical discomfort and increase the need to urinate during the night, disrupting sleep [12].
7. Engaging in regular physical activity: Participating in regular physical activity, preferably during the day or early evening, can help improve sleep quality. However, patients should avoid vigorous exercise close to bedtime, as it may have a stimulating effect [16].

For Long COVID patients with persistent sleep disturbances, cognitive behavioral therapy for insomnia (CBT-I) may be an effective intervention [17]. CBT-I is a structured, short-term therapy that focuses on identifying and changing thoughts and behaviors that interfere with sleep [18]. The main components of CBT-I include sleep education, sleep restriction, stimulus control, cognitive restructuring, and relaxation training [14]. CBT-I has been shown to be effective in improving sleep quality and reducing insomnia severity during the coronavirus disease 19 (COVID-19) pandemic [19].

In addition to sleep hygiene practices and CBT-I, Long COVID patients may benefit from other interventions to improve sleep, such as:

1. Light therapy: Exposure to bright light in the morning can help regulate the circadian rhythm and improve sleep quality [20]. Patients can use a light box or spend time outdoors in natural sunlight.
2. Melatonin supplementation: Melatonin is a hormone that regulates sleep-wake cycles. Some Long COVID patients may benefit from melatonin supplementation, particularly if they have evidence of circadian rhythm disturbances [21]. However, the optimal dosage and timing of melatonin administration remain unclear, and patients should consult with their healthcare provider before starting any supplement regimen.
3. Medications: In some cases, short-term use of sleep medications, such as sedative-hypnotics or low-dose antidepressants with sedating properties, may be considered [22]. However, these medications should be used cautiously and under close medical supervision, as they can have side effects and may not address the underlying causes of sleep disturbances.

Improving sleep quality through proper sleep hygiene practices and targeted interventions is a critical aspect of managing fatigue in Long COVID. Healthcare providers should routinely assess patients' sleep patterns and quality and provide individualized recommendations for improving sleep. A multidisciplinary approach, involving primary care providers, sleep specialists, and mental health professionals, may be necessary for patients with complex or persistent sleep disturbances [1].

Sleep hygiene is an essential component of the non-pharmacological management of fatigue in Long COVID. By promoting healthy sleep habits and addressing sleep disturbances, sleep hygiene can help improve sleep quality, reduce fatigue, and enhance overall functioning in Long COVID patients. CBT-I, light therapy, melatonin supplementation, and medications may be considered for patients with persistent sleep issues but should be used judiciously and in conjunction with sleep hygiene practices. As research on Long COVID continues to evolve, further studies are needed to refine and optimize sleep-based interventions for this patient population.

15.4 Nutrition and Hydration

While the exact mechanisms underlying Long COVID fatigue remain unclear, evidence suggests that nutrition and hydration play a crucial role in managing symptoms and supporting recovery [23]. This chapter will explore the potential benefits of optimizing nutrition and hydration in the treatment of Long COVID fatigue and provide practical recommendations for their implementation.

Nutrition: A balanced and nutrient-dense diet is essential for supporting immune function, reducing inflammation, and promoting overall health in individuals with Long COVID [24]. The following nutritional strategies have been suggested for managing fatigue and other symptoms in Long COVID patients:

1. Adequate protein intake: Consuming sufficient high-quality protein is crucial for maintaining muscle mass, supporting immune function, and promoting tissue repair [25]. Long COVID patients should aim to include lean protein sources, such as poultry, fish, legumes, and low-fat dairy products, in each meal and snack.
2. Anti-inflammatory diet: Adopting an anti-inflammatory diet, rich in fruits, vegetables, whole grains, and healthy fats (e.g., omega-3 fatty acids) may help reduce systemic inflammation and improve fatigue in Long COVID patients [26]. Foods with anti-inflammatory properties include fatty fish, nuts, seeds, olive oil, and colorful fruits and vegetables [27].
3. Adequate vitamin and mineral intake: Micronutrient deficiencies, particularly in vitamins D, C, and B12, as well as zinc and selenium, have been associated with impaired immune function and increased fatigue [28]. Long COVID patients should ensure adequate intake of these nutrients through a balanced diet or, if necessary, through supplementation under medical supervision [29].
4. Probiotics and prebiotics: Emerging evidence suggests that the gut microbiome may play a role in the pathogenesis of Long COVID [30]. Consuming probiotic-rich foods, such as yogurt, kefir, and fermented vegetables, as well as prebiotic fibers found in fruits, vegetables, and whole grains, may help support gut health and potentially reduce fatigue [31].
5. Limiting processed and high-sugar foods: Processed foods and those high in added sugars can contribute to inflammation and exacerbate fatigue [26]. Long COVID patients should limit their intake of these foods and instead focus on whole, minimally processed foods.

A notable clinical trial by Tosato et al. [32] has provided compelling evidence for the efficacy of specific nutritional supplementation in treating Long COVID fatigue.

This study describes a single-blind randomized controlled trial investigating the effects of L-arginine plus vitamin C supplementation in adults with Long COVID. The study, conducted at the Fondazione Policlinico A. Gemelli IRCCS in Rome between July 2021 and April 2022 involved 46 participants aged 20–60 years who were experiencing persistent fatigue following COVID-19 infection.

The researchers divided participants into two groups: one receiving twice-daily oral supplementation of 1.66 g L-arginine plus 500 mg liposomal vitamin C, and the other receiving a placebo, for 28 days. The primary outcome measured was the distance walked during a 6-minute walk test, while secondary outcomes included handgrip strength, flow-mediated dilation (a measure of endothelial function), and fatigue persistence.

The results showed significant improvements in the supplement group compared to the placebo group. Participants taking L-arginine plus vitamin C showed a meaningful increase in their 6-minute walking distance (+30 meters versus no change in the placebo group), greater improvements in handgrip strength, and better flow-mediated dilation. Perhaps most notably, after 28 days, only two participants (8.7%) in the supplement group still reported fatigue compared to 21 participants (80.1%) in the placebo group.

The study also found that participants' baseline serum L-arginine levels were lower than those of healthy controls, but supplementation successfully restored these levels to normal ranges. The treatment was well tolerated with no reported adverse events. Multivariate analysis revealed that the treatment effects were more pronounced in women than men, though this may have been due to the smaller number of male participants.

The researchers concluded that L-arginine plus vitamin C supplementation effectively improved exercise capacity, muscle strength, and endothelial function while reducing fatigue in adults with Long COVID. These benefits appear to work through increasing nitric oxide bioavailability, which plays a crucial role in vascular and muscle function. While the results are promising, the authors acknowledge that larger, multi-center trials would be valuable to confirm these findings and suggest that future studies might benefit from more sophisticated aerobic and anaerobic testing methods.

15.4.1 Medicinal Plants for the Management of Long COVID-19 Fatigue

A comprehensive review by Chen et al. [33] examines the role of medicinal plants in managing Long COVID fatigue.

Among the medicinal plants studied, Bu-Zhong-Yi-Qi Decoction emerges as one of the most significant traditional formulations. According to Tokumasu et al. [34], this decoction was prescribed to 71.6% of patients for managing post-COVID-19 fatigue, making it the most frequently used treatment. Its effectiveness stems from its dual action mechanism: it exhibits strong anti-inflammatory properties by targeting specific proteins including IL-1β and IL6, while also providing antioxidant benefits through enhanced SOD activity and reduction of harmful ROS levels.

Dang-Gui-Shao-Yao San represents another important traditional formulation, with research by Fujiwara et al. [35] demonstrating its comprehensive approach to fighting fatigue. The formula works by modulating inflammatory processes through the reduction of key inflammatory mediators such as IL-2 and IL-1β. Additionally, it provides significant antioxidant protection through free radical scavenging and enhancement of SOD activity.

Ling-Gui-Zhu-Gan Decoction, as studied by Wang et al. [36], focuses primarily on antioxidant mechanisms. The formulation has shown remarkable ability to reduce oxidative stress markers while enhancing the body's natural antioxidant defenses, particularly through SOD activity enhancement and regulation of the oxidant-antioxidant balance.

Qingjin Yiqi Granules have demonstrated particularly promising results in clinical trials. Pang et al. [37] reported significant improvements in fatigue symptoms through enhanced SOD activity and increased expression of key cellular regulators Akt and Nrf2 mRNA in skeletal muscle tissue.

HRG80 Red Ginseng represents a more modern approach to traditional medicine. Teitelbaum et al. [38] documented impressive clinical results, with over 60% of participants reporting recovery from fatigue symptoms. The compound works by improving mitochondrial function and enhancing cellular energy production through multiple mechanisms, including improved muscle metabolism and enhanced enzyme activities.

A novel approach comes in the form of an essential oil blend combining thyme, orange peel, clove bud, and frankincense. While its exact mechanism remains under investigation, clinical trials by Hawkins et al. [39] showed significant reductions in fatigue symptoms when administered through inhalation, as measured by the multidimensional fatigue symptom inventory.

Huangjing Sanqi extract, derived from polygonati rhizoma and notoginseng radix, works through sophisticated cellular pathways. Yang et al. [40] demonstrated its effectiveness in reducing oxidative stress markers while enhancing antioxidant enzyme activities, particularly through the E2F1 and PI3K-AKT pathway interaction.

These medicinal plants represent promising therapeutic options for post-COVID-19 fatigue, though the authors emphasize the need for more extensive clinical research to fully validate their effectiveness and understand their mechanisms of action. Future research should focus on identifying specific active compounds and establishing standardized treatment protocols.

Hydration: Adequate hydration is crucial for maintaining overall health and preventing dehydration, which can exacerbate fatigue and other symptoms in Long COVID patients. The following hydration strategies are recommended:

1. Adequate fluid intake: Long COVID patients should aim to drink at least 8 cups (64 ounces) of water or other non-caffeinated, low-sugar beverages per day [41]. Individual fluid needs may vary based on factors such as age, sex, weight, and activity level.
2. Electrolyte balance: Maintaining proper electrolyte balance, particularly sodium and potassium, is essential for hydration and cellular function [41]. Long COVID patients should consider consuming foods rich in electrolytes, such as fruits, vegetables, and low-fat dairy products, or using electrolyte replacement drinks if necessary.
3. Limiting caffeine and alcohol: Caffeine and alcohol can have diuretic effects, leading to increased fluid loss and potentially exacerbating dehydration and

fatigue [41]. Long COVID patients should limit their intake of these beverages and prioritize water and other hydrating fluids.

Individualized Nutrition and Hydration Plans: Given the heterogeneous nature of Long COVID, nutrition and hydration strategies should be tailored to each patient's unique needs, preferences, and symptoms [1]. Healthcare providers, including registered dietitians, should work closely with Long COVID patients to develop personalized nutrition and hydration plans that consider the following factors:

1. Comorbidities: Long COVID patients with preexisting or concurrent medical conditions, such as diabetes, cardiovascular disease, or gastrointestinal disorders, may require specific dietary modifications to manage their conditions and optimize their overall health [42].
2. Food intolerances and allergies: Some Long COVID patients may develop new or worsening food intolerances or allergies [43]. Identifying and avoiding trigger foods while ensuring adequate nutrient intake is essential for managing symptoms and preventing nutrient deficiencies.
3. Medications and supplements: Long COVID patients may be prescribed various medications or supplements to manage their symptoms [44]. Healthcare providers should consider potential drug-nutrient interactions and adjust nutrition and hydration plans accordingly.
4. Socioeconomic factors: Access to healthy foods and safe drinking water may be limited for some Long COVID patients due to socioeconomic barriers [45]. Healthcare providers should consider these factors and provide resources and support to help patients overcome obstacles to optimal nutrition and hydration.

Monitoring and Adjusting Plans: Regular monitoring and adjustment of nutrition and hydration plans are essential for ensuring their effectiveness and addressing any challenges or concerns that may arise [1]. Healthcare providers should:

1. Assess nutrient intake and hydration status: Regularly evaluate patients' dietary intake and hydration status using tools such as food diaries, 24-hour recalls, and urine color charts [46].
2. Monitor symptoms and quality of life: Assess the impact of nutrition and hydration interventions on fatigue and other Long COVID symptoms, as well as overall quality of life, using validated questionnaires and patient-reported outcomes [47].
3. Adjust plans as needed: Based on the assessment of nutrient intake, hydration status, symptoms, and quality of life, healthcare providers should adjust nutrition and hydration plans as needed to optimize patient outcomes.

Nutrition and hydration are essential components of the comprehensive management of Long COVID fatigue. By adopting a balanced, anti-inflammatory diet, ensuring adequate fluid and electrolyte intake, and tailoring interventions to individual needs and preferences, Long COVID patients may experience improvements in fatigue and overall health. Healthcare providers, including registered dietitians, play a crucial role in developing, monitoring, and adjusting personalized nutrition and hydration plans to support recovery and enhance quality of life in this patient population. As research on Long COVID continues to evolve, further studies are needed to refine and optimize nutrition and hydration strategies for managing fatigue and other symptoms.

15.5 Enzyme and Probiotic Therapy in Long COVID Fatigue

Rathi et al. [48] describe a randomized controlled trial investigating the effectiveness of systemic enzymes (ImmunoSEB) and probiotics (ProbioSEB CSC3) in treating post-COVID-19 fatigue. The study involved 200 patients who had recovered from COVID-19 but were experiencing persistent fatigue. The participants were divided into two groups: 100 received the supplements for 14 days, while 100 received a placebo.

The researchers used the Chalder Fatigue Scale (CFQ-11) to measure both physical and mental fatigue throughout the study period. The results showed significant improvements in the treatment group compared to the placebo group. By day 14, 91% of patients in the treatment group were considered fatigue-free, compared to only 15% in the placebo group. The improvement was evident in both physical symptoms (such as tiredness and muscle weakness) and mental symptoms (including concentration and memory difficulties).

The treatment was well tolerated, with no adverse events reported during the study period. The researchers noted that the supplements appeared to be effective regardless of how long after COVID-19 infection the treatment was started, working for both early and long-term fatigue cases.

The study is particularly significant as it represents one of the first randomized controlled trials addressing post-COVID fatigue, a condition that affects many COVID-19 survivors and can persist for months after initial recovery. The authors suggest that the combination of systemic enzymes and probiotics may work through multiple mechanisms, including anti-inflammatory, antioxidant, and immunomodulatory effects.

While the study showed promising results, the authors acknowledged some limitations, including the relatively short follow-up period of two weeks and the lack of long-term monitoring for potential fatigue recurrence. They recommended further studies to evaluate the treatment's effectiveness in patients with chronic fatigue lasting six months or longer and to investigate the specific biological mechanisms behind the improvements observed.

This groundbreaking study provides strong evidence for the potential of combined enzyme and probiotic supplementation in treating Long COVID fatigue. The high response rate and safety profile suggest this approach could be a valuable addition to the therapeutic arsenal. However, as with many Long COVID treatments, longer-term studies are needed to fully understand the durability of these effects and optimal treatment protocols.

15.6 Stress Management

Stress is a common and significant contributor to fatigue in patients with Long COVID, also known as post-acute sequelae of SARS-CoV-2 infection (PASC) [1]. The ongoing physical, emotional, and social challenges associated with Long COVID can lead to chronic stress, which can exacerbate symptoms and impair overall well-being [5]. Therefore, stress management is an essential component of a comprehensive fatigue management plan for patients with Long COVID. This chapter will explore the role of stress management in treating Long COVID fatigue and provide evidence-based techniques and recommendations for their implementation.

Stress and Long COVID Fatigue: Chronic stress can have a profound impact on the body's physiological processes, including the immune, endocrine, and nervous systems [49]. In patients with Long COVID, stress may contribute to fatigue through several mechanisms:

1. Inflammation: Chronic stress can lead to a persistent state of low-grade inflammation, which has been linked to fatigue and other symptoms in Long COVID [5].
2. Hypothalamic-pituitary-adrenal (HPA) axis dysfunction: Stress can disrupt the normal functioning of the HPA axis, which regulates the body's stress response and energy metabolism [50].
3. Autonomic nervous system imbalance: Stress can lead to an imbalance between the sympathetic and parasympathetic branches of the autonomic nervous system, contributing to fatigue and other symptoms [51].
4. Sleep disturbances: Stress can interfere with sleep quality and quantity, exacerbating fatigue and impairing overall functioning [52].

Stress Management Techniques: Several evidence-based stress management techniques have been shown to be effective in reducing stress and improving fatigue in various chronic conditions, including Long COVID [13]. These techniques include:

1. Relaxation techniques: a. Deep breathing exercises: Deep, diaphragmatic breathing can help activate the parasympathetic nervous system, promoting relaxation and reducing stress [52]. Patients should be instructed to take slow, deep breaths, focusing on expanding the abdomen rather than the chest. b. Progressive muscle relaxation (PMR): PMR involves systematically tensing and relaxing different muscle groups throughout the body, promoting physical and mental relaxation [53]. Patients should be guided through a PMR sequence, typically starting with the feet and progressing up to the head.
2. Mindfulness-based stress reduction (MBSR): MBSR is a structured program that combines mindfulness meditation, body awareness, and yoga to help manage stress and improve overall well-being [54]. MBSR has been shown to be effective in reducing fatigue and improving quality of life in various chronic conditions [55]. Patients can be referred to a qualified MBSR instructor or provided with resources for self-guided practice.

3. Cognitive-behavioral therapy (CBT): CBT is a psychotherapeutic approach that helps patients identify and change maladaptive thoughts and behaviors that contribute to stress and fatigue [56]. CBT has been shown to be effective in managing fatigue and improving overall functioning in patients with Long COVID [57]. Patients should be referred to a qualified CBT therapist for individualized treatment.
4. Lifestyle modifications: a. Regular exercise: Engaging in regular, low-to-moderate intensity exercise can help reduce stress, improve mood, and boost energy levels [58]. Patients should be encouraged to participate in activities they enjoy, such as walking, swimming, or yoga, while being mindful of their individual limitations and pacing themselves to avoid exacerbating fatigue. b. Healthy sleep habits: Establishing a consistent sleep schedule, creating a relaxing bedtime routine, and optimizing the sleep environment can help improve sleep quality and reduce stress [59]. Patients should be provided with education and resources on healthy sleep habits, such as the importance of limiting screen time before bed and maintaining a cool, dark, and quiet bedroom. c. Social support: Connecting with family, friends, and support groups can provide a sense of belonging and reduce feelings of isolation and stress. Patients should be encouraged to maintain social connections, either in-person or virtually, and to seek support from others who understand their experiences with Long COVID.

Individualized Stress Management Plans: Given the heterogeneous nature of Long COVID, stress management interventions should be tailored to each patient's unique needs, preferences, and circumstances [1]. Healthcare providers should work closely with patients to develop personalized stress management plans that consider the following factors:

1. Symptom severity: The intensity and frequency of stress management techniques may need to be adjusted based on the severity of the patient's fatigue and other symptoms [57].
2. Comorbidities: Patients with preexisting or co-occurring mental health conditions, such as depression or anxiety disorders, may require additional support and specialized treatment [5].
3. Personal preferences: Patients should be encouraged to choose stress management techniques that resonate with their interests and values, as this may improve adherence and effectiveness [13].
4. Access to resources: The availability and accessibility of stress management resources, such as qualified therapists or support groups, may vary depending on the patient's location and socioeconomic status [52]. Healthcare providers should consider these factors and provide appropriate referrals and resources.

Monitoring and Adjusting Plans: Regular monitoring and adjustment of stress management plans are essential for ensuring their effectiveness and addressing any challenges or concerns that may arise [1]. Healthcare providers should:

1. Assess stress levels and fatigue: Regularly evaluate patients' perceived stress levels and fatigue severity using validated questionnaires, such as the Perceived Stress Scale (PSS) [60] and the Fatigue Severity Scale (FSS) [61].
2. Monitor adherence and effectiveness: Assess patients' adherence to their stress management plans and the effectiveness of the techniques in reducing stress and fatigue. This can be done through patient self-reporting, as well as objective measures, such as changes in cortisol levels or heart rate variability [49].
3. Adjust plans as needed: Based on the assessment of stress levels, fatigue, adherence, and effectiveness, healthcare providers should adjust stress management plans as needed to optimize patient outcomes. This may involve modifying the frequency or intensity of techniques, exploring alternative approaches, or providing additional support and resources.

Stress management is a crucial component of the comprehensive treatment of Long COVID fatigue. By incorporating evidence-based techniques, such as relaxation exercises, mindfulness-based stress reduction, cognitive-behavioral therapy, and lifestyle modifications, patients can learn to better manage stress and improve their overall well-being. Healthcare providers play a key role in developing individualized stress management plans, monitoring their effectiveness, and providing ongoing support and resources. As research on Long COVID continues to evolve, further studies are needed to refine and optimize stress management interventions for this patient population.

15.7 Hyperbaric Oxygen Therapy (HBOT)

Among emerging therapeutic approaches for Long COVID fatigue, hyperbaric oxygen therapy has shown particular promise through rigorous clinical investigation. HBOT's potential mechanism of action involves increasing oxygen delivery to tissues, modulating inflammatory responses, and promoting neuroplasticity. While preliminary studies suggested possible benefits, definitive evidence from controlled trials was initially lacking. However, recent research has now provided more substantial support for this intervention.

A new randomized, double-blind, sham-controlled clinical trial [62] has demonstrated promising results for hyperbaric oxygen therapy (HBOT) in treating post-COVID-19 condition. The study, which included 73 patients with persistent symptoms for at least three months after COVID-19 infection, compared 40 daily sessions of HBOT (100% oxygen at 2 ATA) to a sham treatment (21% oxygen at 1.03 ATA).

The research revealed significant improvements across multiple domains in patients receiving HBOT. Participants showed enhanced cognitive function, particularly in attention and executive function, though memory improvements were observed in both treatment and control groups, suggesting natural recovery in this area. The therapy also led to meaningful reductions in fatigue, with patients reporting increased energy levels and improved physical limitations. Sleep quality improved notably, and patients experienced fewer psychiatric symptoms, including reduced depression, anxiety, and somatization. Pain interference with daily activities also decreased substantially.

The researchers documented biological changes that corresponded with these clinical improvements. Brain imaging showed increased perfusion in multiple regions and significant microstructural changes in areas associated with cognitive and emotional functions. These physical changes correlated well with the observed symptom improvements.

Importantly, the treatment proved to be safe and well tolerated. The occurrence of side effects was similar between the HBOT and control groups, and no patients needed to discontinue treatment due to adverse effects. While these results are encouraging, the authors note that larger studies will be necessary to optimize patient selection and evaluate long-term outcomes. This trial represents one of the first controlled studies offering a potential therapeutic approach for post-COVID condition, suggesting that HBOT might be a valuable treatment option for these patients.

15.8 Clinical Reference Tables for Long COVID Fatigue Management

The following tables provide evidence-based frameworks for the assessment, monitoring, and management of patients with Long COVID fatigue. Developed from current research and clinical experience, these tables are designed to support clinical decision-making while maintaining the flexibility needed for individualized patient care.

These reference tables serve multiple purposes:

- Provide structured approaches to patient assessment
- Guide treatment selection and monitoring
- Support clinical documentation
- Facilitate consistent care delivery
- Enable systematic outcome tracking

Tables 15.1, 15.2, 15.3, and 15.4 focus on clinical management, progressing from initial assessment through treatment selection to monitoring protocols. These tables reflect the staged approach needed for effective Long COVID fatigue management, emphasizing the importance of comprehensive evaluation and systematic follow-up.

Tables 15.5, 15.6, 15.7, 15.8, and 15.9 present validated assessment tools essential for objective measurement of fatigue and related symptoms. These tools have been selected based on their reliability, validity, and practical utility in Long COVID care. Each tool is presented with guidance for administration and interpretation, allowing clinicians to select the most appropriate measures for their specific clinical context.

Table 15.1 Initial assessment of Long COVID fatigue

Assessment category	Key elements	Action points
Red flags	Severe chest pain/pressure, new-onset shortness of breath, severe dizziness/fainting, new cognitive decline, depression/anxiety crisis	Immediate referral to emergency care
Timeline assessment	Date of COVID infection, vaccination status, fatigue onset, progression pattern	Document chronological progression
Symptom characterization	Daily pattern, triggers, impact on activities, post-exertional malaise	Use standardized fatigue scale

Table 15.2 Severity classification and management approach

Severity level	Symptoms	Functional impact	Management approach
Mild	Can perform daily activities, minimal work impact, no PEM	Minor lifestyle modifications	Activity monitoring, sleep hygiene, regular follow-up
Moderate	Some activities affected, reduced work capacity, mild PEMS	Significant lifestyle adjustments	Pacing strategies, occupational therapy, monthly monitoring
Severe	Most activities affected, unable to work, severe PEM	Major functional impairment	Intensive support, disability management, weekly monitoring

Table 15.3 Treatment selection guide

Symptom	First line	Second line	Third line
Fatigue	Pacing, sleep hygiene, nutrition optimization	Occupational therapy, physical therapy	Consider medications, specialist referral
Sleep disturbance	Sleep hygiene, melatonin	Low-dose antidepressants	Short-term sleep medications
Pain	Acetaminophen, NSAIDs	Gabapentinoids	Low-dose naltrexone
Autonomic symptoms	Salt/fluid intake, compression garments	Beta-blockers	Fludrocortisone

Table 15.4 Monitoring and follow-up schedule

Severity	Initial follow-up	Subsequent visits	Monitoring parameters
Mild	4–6 weeks	Every 2–3 months	Fatigue scores, activity levels, work participation
Moderate	2–3 weeks	Monthly	Above plus: Symptom diary, treatment response
Severe	Weekly	Every 2–4 weeks	Above plus: Symptom diary, treatment response

Table 15.5 Validated fatigue assessment scales

Scale name	Number of items	Scoring range	Key features	Best used for
Fatigue severity scale (FSS)	9 items	1–7 per item; mean score calculated	Measures impact on daily function, good validation in post-viral conditions, simple to administer	Initial assessment, Progress monitoring, research studies
Chalder fatigue scale (CFS)	11 items	0–3 per item (Likert) or 0–1 (bimodal)	Physical and mental fatigue subscales, sensitive to change, widely used in long COVID	Detailed fatigue characterization, treatment response monitoring
Modified fatigue impact scale (MFIS)	21 items	0–4 per item; total 0–84	Physical, cognitive, psychosocial domains, comprehensive assessment, good reliability	Detailed impact assessment, research purposes
Fatigue assessment scale (FAS)	10 items	1–5 per item; total 10–50	Physical and mental fatigue, quick to complete, good psychometric properties	Rapid screening, routine monitoring

Table 15.6 Functional assessment tools

Assessment tool	Purpose	Administration time	Key measures	Clinical utility
6-minute walk test	Physical capacity	6 minutes	Distance walked, heart rate, oxygen saturation, perceived exertion	Exercise capacity. Deconditioning, treatment response
Hand grip strength	Muscle strength	5 minutes	Maximum force, fatigue ratio	Physical function, prognosis indicator
Sit-to-stand test	Functional strength	3 minutes	Number of repetitions, time to complete	Lower body strength, fall risk
Short physical performance battery	Overall function	10–15 minutes	Balance, gait speed, chair stand	Comprehensive function, disability risk

When using these tables, clinicians should:

- Consider each patient's individual circumstances and comorbidities
- Adapt approaches based on available resources and local guidelines
- Document modifications to standard protocols
- Monitor treatment responses systematically
- Adjust interventions based on patient progress

The tables are intended to complement, not replace, clinical judgment. They should be used in conjunction with the detailed information provided throughout this chapter and updated as new evidence emerges in this rapidly evolving field.

Table 15.7 Cognitive assessment tools

Tool name	Domains assessed	Time to complete	Scoring	Clinical application
Montreal cognitive assessment (MoCA)	Memory, attention, executive function	10–15 minutes	0–30 points	Cognitive screening, progress monitoring
Trail making test	Processing speed, executive function	5–10 minutes	Time to complete	Attention, cognitive fatigue
Symbol digit modalities test	Information processing, attention	5 minutes	Number correct in 90 seconds	Cognitive function, mental fatigue
Digit span test	Working memory, attention	5–10 minutes	Forward and backward span	Cognitive endurance, mental fatigue

Table 15.8 Quality of life and symptom impact tools

Assessment tool	Focus areas	Format	Time to complete	Use case
SF-36	Physical function, mental health, social function, role limitations	36 questions	10 minutes	Comprehensive QoL, research studies
EQ-5D-5L	Mobility, self-care, usual activities, pain/ discomfort, anxiety/ depression	5 questions + VAS	5 minutes	Quick QoL screen, health economics
Post-COVID functional scale	Daily activities, work capacity, social function	1 item	2 minutes	Rapid assessment, progress tracking
Work productivity scale	Work attendance, productivity, activity impairment	6 questions	5 minutes	Occupational impact, disability assessment

Table 15.9 Autonomic function assessment

Test	Purpose	Parameters measured	Interpretation
Active stand test	Orthostatic response	Blood pressure, heart rate, symptoms	POTS diagnosis, OH assessment
Heart rate variability	Autonomic balance	Time domain measures, frequency domain measures	Autonomic function, treatment monitoring
Composite autonomic severity score	Overall autonomic function	Sudomotor, cardiovascular, adrenergic	Comprehensive evaluation, research use

The clinical reference tables serve as a structured framework while emphasizing the essential role of physician judgment in patient care. Individual patient circumstances, including specific comorbidities, severity of symptoms, and personal factors, must guide the application of these standardized protocols. A patient's access to healthcare resources, local treatment guidelines, and available specialist support significantly influence treatment decisions and monitoring strategies.

Systematic documentation through standardized assessment tools enables tracking of patient progress and treatment effectiveness. Regular monitoring allows clinicians to identify patterns in symptom progression and treatment response, supporting evidence-based adjustments to care plans. Treatment modifications should reflect both objective measurements and patient-reported outcomes, ensuring that interventions remain aligned with individual patient needs and goals.

The dynamic nature of Long COVID research demands flexibility in clinical protocols. As new evidence emerges about disease mechanisms and treatment effectiveness, care guidelines must evolve accordingly. The framework therefore provides a foundation for standardized care while allowing incorporation of new findings and adaptation to individual patient circumstances. This balance between structured guidance and personalized care ensures that treatment remains both systematic and patient-centered, acknowledging the complex and evolving nature of Long COVID fatigue management.

Summary

This comprehensive chapter explores four key non-pharmacological approaches to managing Long COVID fatigue: lifestyle modifications, sleep hygiene, nutrition and hydration, and stress management. The chapter also provides detailed clinical reference tables for standardized assessment and monitoring.

The lifestyle modifications section emphasizes pacing and energy conservation strategies, detailing how patients should balance activity with rest to prevent post-exertional malaise. Key recommendations include prioritizing essential activities, breaking tasks into manageable steps, taking regular breaks, and monitoring energy levels.

Sleep hygiene is addressed through specific interventions such as maintaining consistent sleep schedules, creating optimal sleep environments, and limiting electronic device use before bedtime. The chapter discusses the potential benefits of cognitive behavioral therapy for insomnia (CBT-I) and other interventions like light therapy and melatonin supplementation for patients with persistent sleep disturbances.

The nutrition and hydration section outlines strategies including adequate protein intake, anti-inflammatory diet principles, medical plants, and proper hydration guidelines. It emphasizes the importance of individualizing nutritional approaches based on patient-specific factors, comorbidities, and potential food intolerances.

Stress management techniques are detailed, including relaxation methods, mindfulness-based stress reduction, and cognitive-behavioral therapy. The chapter explores the physiological connections between stress and fatigue in

Long COVID patients, providing evidence-based approaches for stress reduction.

The chapter concludes with nine comprehensive clinical reference tables that provide structured frameworks for assessment, monitoring, and management of Long COVID fatigue. These tables cover various aspects from initial assessment and severity classification to validated assessment tools for fatigue, functional capacity, cognitive function, quality of life, and autonomic function.

Throughout, the chapter emphasizes the importance of individualizing approaches while maintaining systematic assessment and monitoring protocols, acknowledging the evolving nature of Long COVID research and treatment guidelines.

References

1. Nalbandian A, Sehgal K, Gupta A, Madhavan MV, McGroder C, Stevens JS, et al. Post-acute COVID-19 syndrome. Nat Med. 2021;27(4):601–15.
2. Gorna R, MacDermott N, Rayner C, O'Hara M, Evans S, Agyen L, et al. Long COVID guidelines need to reflect lived experience. Lancet. 2021;397(10273):455–7.
3. Humphreys H, Kilby L, Kudiersky N, Copeland R. Long COVID and the role of physical activity: a qualitative study. BMJ Open. 2021;11(3):e047632.
4. Goudsmit EM, Nijs J, Jason LA, Wallman KE. Pacing as a strategy to improve energy management in myalgic encephalomyelitis/chronic fatigue syndrome: a consensus document. Disabil Rehabil. 2012;34(13):1140–7.
5. Ceban F, Ling S, Lui LM, Lee Y, Gill H, Teopiz KM, et al. Fatigue and cognitive impairment in post-COVID-19 syndrome: a systematic review and meta-analysis. Brain Behav Immun. 2022;101:93–135.
6. White PD, Goldsmith KA, Johnson AL, Potts L, Walwyn R, DeCesare JC, et al. Comparison of adaptive pacing therapy, cognitive behavior therapy, graded exercise therapy, and specialist medical care for chronic fatigue syndrome (PACE): a randomized trial. Lancet. 2011;377(9768):823–36.
7. Jason LA, Brown M, Brown A. Energy conservation/envelope theory interventions to help patients with myalgic encephalomyelitis/chronic fatigue syndrome. Fatigue. 2013;1(1–2):27–42.
8. Nijs J, Paul van Wilgen C, Van Oosterwijck J, van Ittersum M, Meeus M. How to explain central sensitization to patients with 'unexplained' chronic musculoskeletal pain: practice guidelines. Man Ther. 2011;16(5):413–8.
9. Jamieson-Lega K, Berry R, Brown CA. Pacing: a concept analysis of the chronic pain intervention. Pain Res Manag. 2013;18(4):207–13.
10. Kuut TA, Müller F, Csorba I, Braamse A, Aldenkamp A, Appelman B, Assmann-Schuilwerve E, Geerlings SE, Gibney KB, Kanaan RAA, Mooij-Kalverda K, Olde Hartman TC, Pauëlsen D, Prins M, Slieker K, van Vugt M, Keijmel SP, Nieuwkerk P, Rovers CP, Knoop H. Efficacy of cognitive-behavioral therapy targeting severe fatigue following coronavirus disease 2019: results of a randomized controlled trial. Clin Infect Dis. 2023;77(5):687–95. https://doi.org/10.1093/cid/ciad257. PMID: 37155736; PMCID: PMC10495128.
11. Huang C, Huang L, Wang Y, Li X, Ren L, Gu X, et al. 6-month consequences of COVID-19 in patients discharged from hospital: a cohort study. Lancet. 2021;397(10270):220–32.

12. Irish LA, Kline CE, Gunn HE, Buysse DJ, Hall MH. The role of sleep hygiene in promoting public health: a review of empirical evidence. Sleep Med Rev. 2015;22:23–36.
13. Reddy S, Reddy V, Sharma S. Physiology, Circadian Rhythm [Internet]. StatPearls. Treasure Island: StatPearls Publishing; 2025 [updated 2023 May 1]. Available from: https://www.ncbi.nlm.nih.gov/books/NBK519507/
14. Shechter A, Kim EW, St-Onge MP, Westwood AJ. Blocking nocturnal blue light for insomnia: a randomized controlled trial. J Psychiatr Res. 2018;96:196–202.
15. Murawski B, Wade L, Plotnikoff RC, Lubans DR, Duncan MJ. A systematic review and meta-analysis of cognitive and behavioral interventions to improve sleep health in adults without sleep disorders. Sleep Med Rev. 2018;40:160–9.
16. Reid KJ, Baron KG, Lu B, Naylor E, Wolfe L, Zee PC. Aerobic exercise improves self-reported sleep and quality of life in older adults with insomnia. Sleep Med. 2010;11(9):934–40.
17. Pedrozo-Pupo JC, Caballero-Domínguez CC, Campo-Arias A. Prevalence and variables associated with insomnia among COVID-19 survivors in Colombia. Acta Biomed. 2022;93(1):e2022019.
18. Trauer JM, Qian MY, Doyle JS, Rajaratnam SM, Cunnington D. Cognitive behavioral therapy for chronic insomnia: a systematic review and meta-analysis. Ann Intern Med. 2015;163(3):191–204.
19. Cheng P, Casement MD, Kalmbach DA, Castelan AC, Drake CL. Digital cognitive behavioral therapy for insomnia promotes later health resilience during the coronavirus disease 19 (COVID-19) pandemic. Sleep. 2021;44(4):zsaa258.
20. Sletten TL, Raman B, Magee M, Ferguson SA, Kennaway DJ, Grunstein RR, et al. A blue-enriched, increased intensity light intervention to improve alertness and performance in rotating night shift workers in an operational setting. Nat Sci Sleep. 2021;13:647–57.
21. Cardinali DP, Brown GM, Pandi-Perumal SR. Possible application of melatonin in long COVID. Biomol Ther. 2022;12(11):1646.
22. Fenton C, Lee A. Antidepressants with anti-inflammatory properties may be useful in long COVID depression. Drugs Ther Perspect. 2023;39(2):65–70.
23. Barrea L, Muscogiuri G, Frias-Toral E, Laudisio D, Pugliese G, Castellucci B, et al. Nutrition and immune system: from the Mediterranean diet to dietary supplementary through the microbiota. Crit Rev Food Sci Nutr. 2020;61(18):1–25.
24. Iddir M, Brito A, Dingeo G, Fernandez Del Campo SS, Samouda H, La Frano MR, et al. Strengthening the immune system and reducing inflammation and oxidative stress through diet and nutrition: considerations during the COVID-19 crisis. Nutrients. 2020;12(6):1562.
25. Jayawardena R, Sooriyaarachchi P, Chourdakis M, Jeewandara C, Ranasinghe P. Enhancing immunity in viral infections, with special emphasis on COVID-19: a review. Diabetes Metab Syndr. 2020;14(4):367–82.
26. Zabetakis I, Lordan R, Norton C, Tsoupras A. COVID-19: the inflammation link and the role of nutrition in potential mitigation. Nutrients. 2020;12(5):1466.
27. Ling V, Zabetakis I. The role of an anti-inflammatory diet in conjunction to COVID-19. Diseases. 2021;9(4):76.
28. Shakoor H, Feehan J, Dhaheri ASA, Ali HI, Platat C, Ismail LC, et al. Immune-boosting role of vitamins D, C, E, zinc, selenium and omega-3 fatty acids: could they help against COVID-19? Maturitas. 2021;143:1–9.
29. Alexander J, Tinkov A, Strand TA, Alehagen U, Skalny A, Aaseth J. Early nutritional interventions with zinc, selenium and vitamin D for raising anti-viral resistance against progressive COVID-19. Nutrients. 2020;12(8):2358.
30. Nguyen QV, Chong LC, Hor YY, Lew LC, Rather IA, Choi SB. Role of probiotics in the management of COVID-19: a computational perspective. Nutrients. 2022;14:274.
31. Alenazy MF, Aljohar HI, Alruwaili AR, Daghestani MH, Alonazi MA, Labban RS, et al. Gut microbiota dynamics in relation to long-COVID-19 syndrome: role of probiotics to combat psychiatric complications. Meta. 2022;12(10):912.
32. Tosato M, Calvani R, Picca A, Ciciarello F, Galluzzo V, Coelho-Júnior HJ, Di Giorgio A, Di Mario C, Gervasoni J, Gremese E, Leone PM, Nesci A, Paglionico AM, Santoliquido A,

Santoro L, Santucci L, Tolusso B, Urbani A, Marini F, Marzetti E, Landi F; Gemelli against COVID-19 Post-Acute Care Team. Effects of l-Arginine plus vitamin C supplementation on physical performance, endothelial function, and persistent fatigue in adults with long COVID: A single-blind randomized controlled trial. Nutrients. 2022;14(23):4984. https://doi.org/10.3390/nu14234984. PMID: 36501014; PMCID: PMC9738241.
33. Chen Y, Zhang C, Feng Y. Medicinal plants for the management of post-COVID-19 fatigue: a literature review on the role and mechanisms. J Tradit Complement Med. 2024;15(1):15–23.
34. Tokumasu K, Ueda K, Honda H, Terada K, Tsumura K. Application of kampo medicines for treatment of general fatigue due to long COVID. Medicina (Kaunas). 2022;58(6):730.
35. Fujiwara M, Egashira N, Mishima K, Iwasaki K. Tokishakuyakusan: pharmacological evidence for neuroactive and neuroprotective actions. J Tradit Med. 2007;24(5):149–55.
36. Wang ZF, Hu YQ, Wu QG, Zhang R. Virtual screening of potential anti-fatigue mechanism of Polygonati rhizoma based on network pharmacology. Comb Chem High Throughput Screen. 2019;22(9):612–24.
37. Pang W, Yang F, Zhao Y, Liu X, Wang G, Chen L, et al. Qingjin Yiqi granules for post-COVID-19 condition: a randomized clinical trial. J Evid Based Med. 2022;15(1):30–8.
38. Teitelbaum J, Goudie S. An open-label, pilot trial of HRG80 red ginseng in chronic fatigue syndrome, fibromyalgia, and post-viral fatigue. Pharmaceuticals. 2022;15(1):43.
39. Hawkins J, Hires C, Keenan L, Dunne E. Aromatherapy blend of thyme, orange, clove bud, and frankincense boosts energy levels in post-COVID-19 female patients: a randomized, double-blinded, placebo controlled clinical trial. Complement Ther Med. 2022;67:102823.
40. Yang YQ, Li YQ, Yu LP, Wang XY, Zheng X, Yu Y, et al. Muscle fatigue-alleviating effects of a prescription composed of Polygonati rhizoma and notoginseng radix et rhizoma. Biomed Res Int. 2020;2020:3963045.
41. Popkin BM, D'Anci KE, Rosenberg IH. Water, hydration, and health. Nutr Rev. 2010;68(8):439–58.
42. Mechanick JI, Apovian C, Brethauer S, Garvey WT, Joffe AM, Kim J, et al. Clinical practice guidelines for the perioperative nutrition, metabolic, and nonsurgical support of patients undergoing bariatric procedures - 2019 update. Surg Obes Relat Dis. 2020;16(2):175–247.
43. Trogen B, Jin H, Cianferoni A, Chehade M, Schultz F, Chavez A, et al. A survey examining the impact of COVID-19 on food protein-induced enterocolitis syndrome. J Allergy Clin Immunol Pract. 2022;10(1):312–4.
44. Crook H, Raza S, Nowell J, Young M, Edison P. Long COVID—mechanisms, risk factors, and management. BMJ. 2021;374:n1648.
45. Lukkahatai N, Rodney T, Ling C, Daniel B, Han HR. Long COVID in the context of social determinants of health. Front Public Health. 2023;11:1098443.
46. Armstrong LE. Assessing hydration status: the elusive gold standard. J Am Coll Nutr. 2007;26(5 Suppl):575S–84S.
47. Santus P, Tursi F, Croce G, Di Simone C, Frassanito F, Gaboardi P, et al. Changes in quality of life and dyspnoea after hospitalization in COVID-19 patients discharged at home. Multidiscip Respir Med. 2020;15(1):713.
48. Rathi A, Jadhav SB, Shah N. A randomized controlled trial of the efficacy of systemic enzymes and probiotics in the resolution of post-COVID fatigue. Medicines (Basel). 2021;8(9):47. https://doi.org/10.3390/medicines8090047. PMID: 34564089; PMCID: PMC8472462.
49. Yaribeygi H, Panahi Y, Sahraei H, Johnston TP, Sahebkar A. The impact of stress on body function: a review. EXCLI J. 2017;16:1057–72.
50. Durcan E, Hacioglu A, Karaca Z, Unluhizarci K, Gonen MS, Kelestimur F. Hypothalamic-pituitary Axis function and adrenal insufficiency in COVID-19 patients. Neuroimmunomodulation. 2023;30(1):215–25.
51. Yong SJ. Persistent brainstem dysfunction in long-COVID: a hypothesis. ACS Chem Neurosci. 2021;12(4):573–80.
52. Ma X, Yue ZQ, Gong ZQ, Zhang H, Duan NY, Shi YT, et al. The effect of diaphragmatic breathing on attention, negative effects and stress in healthy adults. Front Psychol. 2017;8:874.

53. Dolbier CL, Rush TE. Efficacy of abbreviated progressive muscle relaxation in a high-stress college sample. Int J Stress Manag. 2012;19(1):48–68.
54. Kabat-Zinn J. Full catastrophe living: using the wisdom of your body and mind to face stress, pain, and illness. Revised ed. New York: Bantam; 2013.
55. Potes A, Souza G, Nikolitch K, Penheiro R, Moussa Y, Jarvis E, et al. Mindfulness in severe and persistent mental illness: a systematic review. Int J Psychiatry Clin Pract. 2018;22(4):253–61.
56. Beck JS. Cognitive behavior therapy: basics and beyond. 2nd ed. New York: Guilford Press; 2011.
57. Crandall A, Daines C, Hanson CL, Barnes MD. The effects of COVID-19 stressors and family life on anxiety and depression one-year into the COVID-19 pandemic. Fam Process. 2023;62(1):336–51.
58. Stanton R, To QG, Khalesi S, Williams SL, Alley SJ, Thwaite TL, et al. Depression, anxiety and stress during COVID-19: associations with changes in physical activity, sleep, tobacco and alcohol use in Australian adults. Int J Environ Res Public Health. 2020;17(11):4065.
59. Morin CM, Vézina-Im LA, Ivers H, Micoulaud-Franchi JA, Philip P, Lamy M, et al. Prevalent, incident, and persistent insomnia in a population-based cohort tested before (2018) and during the first-wave of COVID-19 pandemic (2020). Sleep. 2022;45(1):zsab258.
60. Cohen S, Kamarck T, Mermelstein R. A global measure of perceived stress. J Health Soc Behav. 1983;24(4):385–96.
61. Krupp LB, LaRocca NG, Muir-Nash J, Steinberg AD. The fatigue severity scale: application to patients with multiple sclerosis and systemic lupus erythematosus. Arch Neurol. 1989;46(10):1121–3.
62. Zilberman-Itskovich S, Catalogna M, Sasson E, Elman-Shina K, Hadanny A, Lang E, Finci S, Polak N, Fishlev G, Korin C, Shorer R, Parag Y, Sova M, Efrati S. Hyperbaric oxygen therapy improves neurocognitive functions and symptoms of post-COVID condition: randomized controlled trial. Sci Rep. 2022;12(1):11252. https://doi.org/10.1038/s41598-022-15565-0. PMID: 35821512; PMCID: PMC9276805.

Pharmacological and Non-pharmacological Interventions 16

16.1 Pharmacologic Treatments: Stimulants

While non-pharmacologic interventions, such as pacing, cognitive-behavioral therapy, and graded exercise therapy, remain the first-line treatments for managing fatigue in Long COVID, some patients may require additional pharmacologic support [1]. This chapter will explore the potential role of stimulants in treating Long COVID fatigue and provide an overview of the current evidence and recommendations for their use.

Stimulants and Long COVID Fatigue: Stimulants, such as methylphenidate and modafinil, are medications that increase alertness, attention, and energy by enhancing the activity of certain neurotransmitters in the brain, particularly dopamine and norepinephrine [2]. These medications have been used off-label to treat fatigue in various chronic conditions, such as multiple sclerosis (MS), Parkinson's disease (PD), and chronic fatigue syndrome (CFS) [3, 4].

The rationale for using stimulants in Long COVID fatigue is based on the hypothesis that the persistent fatigue may be related to dysregulation of the central nervous system, particularly the dopaminergic and noradrenergic pathways [5]. By enhancing the activity of these neurotransmitters, stimulants may help alleviate fatigue and improve overall functioning in some patients with Long COVID.

Current Evidence: To date, there is limited evidence specifically evaluating the efficacy and safety of stimulants in treating Long COVID fatigue. Most of the available data come from small case series, observational studies, and extrapolation from other chronic conditions with similar fatigue presentations.

1. Methylphenidate: Methylphenidate is a central nervous system stimulant commonly used to treat attention-deficit/hyperactivity disorder (ADHD) [2]. A case series by Bhattacharjee et al. [6] reported improvements in fatigue and cognitive function in three patients with Long COVID treated with low-dose methylphenidate (10–20 mg/day). However, larger controlled trials are needed to confirm

T. Rudroff, *Long COVID Fatigue*, https://doi.org/10.1007/978-3-031-89474-9_16

these findings and assess the long-term safety and efficacy of methylphenidate in this population.

2. Modafinil: Modafinil is a wakefulness-promoting agent approved for the treatment of narcolepsy, shift work sleep disorder, and excessive daytime sleepiness associated with obstructive sleep apnea (OSA) [7]. In a retrospective cohort study, Miskowiak et al. [8] found that modafinil (200–400 mg/day) was associated with improvements in fatigue and cognitive function in patients with post-COVID-19 syndrome. However, the study was limited by its retrospective design and lack of a control group.
3. Evidence from other chronic conditions: Several randomized controlled trials have investigated the use of stimulants for fatigue in other chronic conditions, with mixed results. For example, a meta-analysis by Zifko et al. [7] found that methylphenidate and modafinil were associated with small but significant improvements in fatigue in patients with MS. However, the authors noted that the quality of evidence was low, and the clinical significance of the observed effects was uncertain. Recommendations and Precautions: Given the limited and inconclusive evidence, the use of stimulants for treating Long COVID fatigue should be considered cautiously and on a case-by-case basis [1]. Healthcare providers should consider the following factors when deciding whether to prescribe stimulants for Long COVID fatigue:
   1. Severity and impact of fatigue: Stimulants may be considered for patients with severe, persistent fatigue that significantly impairs daily functioning and quality of life, and who have not responded to non-pharmacologic interventions [6].
   2. Comorbidities and contraindications: Stimulants should be used with caution in patients with preexisting cardiovascular disease, hypertension, arrhythmias, or a history of substance abuse [2]. They should also be avoided in patients with uncontrolled anxiety, psychosis, or bipolar disorder.
   3. Potential side effects: Common side effects of stimulants include insomnia, headache, dry mouth, decreased appetite, and increased heart rate and blood pressure [7]. Patients should be closely monitored for these side effects and the medication should be discontinued if the risks outweigh the benefits.
   4. Drug interactions: Stimulants may interact with other medications, such as antidepressants, antihypertensives, and anticonvulsants [2]. Healthcare providers should review the patient's medication list and adjust dosages as needed to minimize the risk of drug interactions.
   5. Long-term safety and efficacy: The long-term safety and efficacy of stimulants in treating Long COVID fatigue are unknown [9]. Patients should be informed of the potential risks and benefits of long-term use, and the need for periodic re-evaluation of the treatment plan.

If a decision is made to trial a stimulant for Long COVID fatigue, it is recommended to start with a low dose and titrate slowly based on response and tolerability [6]. The medication should be discontinued if there is no significant improvement in fatigue after an adequate trial (e.g., 4–6 weeks) or if the side effects are intolerable.

Stimulants, such as methylphenidate and modafinil, may have a potential role in treating severe, persistent fatigue in some patients with Long COVID who have not responded to non-pharmacologic interventions. However, the current evidence supporting their use is limited and inconclusive, and their long-term safety and efficacy in this population are unknown. Healthcare providers should carefully consider the potential risks and benefits of stimulants on a case-by-case basis, considering the patient's comorbidities, contraindications, and medication list. If prescribed, stimulants should be started at a low dose, titrated slowly, and discontinued if there is no significant improvement or if the side effects are intolerable. Further research, including randomized controlled trials, is needed to better understand the role of stimulants in managing Long COVID fatigue and to guide evidence-based recommendations for their use.

16.2 Other Symptom-Focused Medications

In addition to stimulants, other symptom-focused medications may be considered for the management of fatigue and associated symptoms in patients with Long COVID, also known as post-acute sequelae of SARS-CoV-2 infection (PASC) [9]. These medications target specific symptoms or underlying mechanisms that may contribute to fatigue, such as pain, autonomic dysfunction, or cognitive impairment [10]. This chapter will provide an overview of other symptom-focused medications that may be used to treat Long COVID fatigue, along with the current evidence and recommendations for their use.

1. Pain Management: Chronic pain is a common symptom in patients with Long COVID and can significantly contribute to fatigue and reduced quality of life [11]. Depending on the type and severity of pain, various medications may be prescribed, including the following: Non-steroidal anti-inflammatory drugs (NSAIDs): Lee et al. (2024) examined the association between nonsteroidal anti-inflammatory drug (NSAID) use and the development of Long COVID using data from the Korean National Health Insurance Service. The study analyzed 225,458 COVID-19 patients diagnosed between 2020 and 2021, of whom 254 developed Long COVID. The researchers found that NSAID use during acute COVID-19 (first 14 days after diagnosis) was associated with higher odds of developing Long COVID (adjusted OR 1.79). However, this association became statistically non-significant when accounting for self-quarantine duration. Interestingly, NSAID use during the chronic phase (after 14 days) was associated with lower odds of Long COVID (adjusted OR 0.54). For comparison, acetaminophen use showed no significant association during the acute phase but was associated with lower odds of Long COVID during the chronic phase. Antiviral drug use showed a strong association with increased odds of Long COVID (adjusted OR 3.75). Acetaminophen is a pain reliever and fever reducer that can be used to manage mild to moderate pain (Schilling et al., 2010). It is generally well tolerated but should be used with caution in patients with liver dysfunction [12]

The study suggests that while NSAID use in acute COVID-19 may increase Long COVID risk, chronic NSAID use might have a protective effect. However, the researchers note several limitations, including the short study timeframe, lack of data on COVID-19 variants, and potential underdiagnosis of Long COVID cases. They emphasize the need for further research to understand the biological mechanisms behind these contrasting effects and to establish clearer causal associations between treatment choices and Long COVID development.

Gabapentinoids: Gabapentin and pregabalin are medications primarily used to treat neuropathic pain, which is caused by damage or dysfunction of the nerves [13]. These medications may be helpful for patients with Long COVID who experience neuropathic pain, such as burning, tingling, or numbness [14].

Low-dose naltrexone (LDN): Naltrexone is an opioid antagonist typically used to treat alcohol and opioid dependence [15]. However, low-dose naltrexone (LDN) has been suggested as a potential treatment for chronic pain and fatigue in various conditions, including fibromyalgia and chronic fatigue syndrome (CFS) [16]. The exact mechanisms by which LDN may help reduce pain and fatigue are not fully understood, but it is thought to modulate the immune system and reduce inflammation [17].

Isman et al. [18], published in Brain, Behavior, & Immunity—Health in 2024, investigated the potential benefits of low-dose naltrexone (LDN) and NAD+ supplementation for treating persistent fatigue symptoms in post-COVID-19 patients. The researchers conducted an observational open-label pilot study involving 36 patients who experienced moderate to severe fatigue 1–12 months after COVID-19 infection. Participants received 4.5 mg/day of LDN and weekly NAD+ supplementation through iontophoresis patches for 12 weeks.

The study found significant improvements in quality of life measures using the SF-36 survey, with mean total scores increasing from 36.5 at baseline to 52.1 after 12 weeks of treatment. Participants also showed marked improvement in fatigue symptoms as measured by the Chalder Fatigue Scale, with scores decreasing from 25.9 at baseline to 17.4 at week 12. Notably, 52% of patients were classified as responders to the treatment, defined as those showing greater than 20% improvement in SF-36 scores and achieving a week-12 score of ≥55.

The treatment was generally well tolerated, with mild adverse events primarily associated with LDN that could be managed through dose adjustments. The most common side effect was skin irritation at the NAD+ patch application site, affecting 25% of patients but resolving quickly. Five participants withdrew from the study, with three citing preexisting medical issues or adverse events known to be caused by LDN.

While the study showed promising results, it had several limitations, including the lack of a control group due to recruitment challenges and its observational nature. The researchers noted that many patients were actively seeking treatment and were unwilling to participate in a placebo-controlled trial. Additionally, the study relied on subjective survey measures rather than objective health measurements.

The findings suggest that combination therapy with LDN and NAD+ may benefit some patients with persistent post-COVID fatigue, though the researchers emphasize the need for larger randomized controlled trials to confirm these results and identify which patient subpopulations might benefit most from this treatment approach. This research contributes to the growing body of evidence for potential therapeutic interventions in Long COVID, while acknowledging the need for further investigation to establish optimal treatment protocols.

2. Autonomic Dysfunction Management: Autonomic dysfunction, which involves the dysregulation of the autonomic nervous system, is a common feature of Long COVID and can manifest as symptoms such as dizziness, lightheadedness, palpitations, and orthostatic intolerance [19]. Medications that may be used to manage autonomic dysfunction in Long COVID include the following: (a) Beta-blockers: Beta-blockers, such as propranolol or metoprolol, can help regulate heart rate and blood pressure, which may be beneficial for patients with palpitations or orthostatic intolerance [20]. (b) Midodrine: Midodrine is a vasoconstrictor that helps increase blood pressure and reduce symptoms of orthostatic intolerance [21]. It may be prescribed for patients with postural orthostatic tachycardia syndrome (POTS) or other forms of orthostatic intolerance [22]. (c) Fludrocortisone: Fludrocortisone is a mineralocorticoid that helps increase blood volume and improve blood pressure regulation [23]. It may be prescribed for patients with orthostatic intolerance or POTS [24].
3. Cognitive Impairment Management: Cognitive impairment, often referred to as "brain fog," is another common symptom in Long COVID and can contribute to fatigue and reduced quality of life [25]. Medications that may be used to manage cognitive impairment in Long COVID include the following (a). Acetylcholinesterase inhibitors: Acetylcholinesterase inhibitors, such as donepezil or rivastigmine, are primarily used to treat Alzheimer's disease and other forms of dementia [26]. They work by increasing the levels of acetylcholine, a neurotransmitter involved in memory and learning [27]. These medications may be considered for patients with Long COVID who have significant cognitive impairment, but their efficacy and safety in this population have not been well established [28]. (b). Modafinil: As discussed in the previous chapter on stimulants, modafinil is a wakefulness-promoting agent that may help improve alertness, concentration, and overall cognitive function [7]. However, its efficacy and safety in treating cognitive impairment in Long COVID have not been specifically studied [5].

Recommendations and Precautions: The use of symptom-focused medications in patients with Long COVID should be guided by a thorough assessment of the patient's individual symptoms, comorbidities, and potential contraindications [9]. Healthcare providers should consider the following factors when prescribing these medications:

1. Potential side effects: Each medication has its own set of potential side effects, which should be carefully considered and monitored [10]. Patients should be

informed of the potential risks and benefits of each medication and encouraged to report any adverse effects to their healthcare provider.
2. Drug interactions: Many symptom-focused medications can interact with other medications, potentially altering their efficacy or increasing the risk of side effects. Healthcare providers should review the patient's medication list and adjust dosages as needed to minimize the risk of drug interactions.
3. Comorbidities and contraindications: Patients with preexisting medical conditions, such as cardiovascular disease, liver or kidney dysfunction, or neurological disorders, may be at increased risk of adverse effects from certain medications [14]. Healthcare providers should carefully consider these comorbidities and potential contraindications when selecting appropriate medications.
4. Limited evidence: The efficacy and safety of many symptom-focused medications in treating Long COVID fatigue have not been well established, and the available evidence is often limited to small studies or case reports [9]. Patients should be informed of the limitations of the current evidence and the need for further research to guide evidence-based recommendations.

16.3 Amantadine for Long COVID Fatigue

Among emerging pharmacological treatments for Long COVID fatigue, Amantadine has attracted particular interest due to its established use in treating fatigue associated with multiple sclerosis and other neurological conditions. Recent clinical research has begun to evaluate its potential effectiveness specifically for post-COVID fatigue.

An open-label randomized clinical trial by Harandi et al. [29] investigated the effectiveness of Amantadine for treating post-COVID-19 fatigue. The study included 66 patients who experienced persistent fatigue 30–60 days after confirmed COVID-19 infection. The research team divided participants into two groups: an intervention group receiving 100 mg Amantadine twice daily for two weeks, and a control group receiving no treatment.

The researchers evaluated fatigue levels using two measurement tools: the Visual Analog Fatigue Scale (VAFS) and the Fatigue Severity Scale (FSS). At the start of the study, both groups showed similar baseline fatigue scores. After two weeks of treatment, the Amantadine group demonstrated significant improvement compared to the control group. The VAFS scores in the Amantadine group decreased from 7.90 to 3.37, while the control group showed a smaller reduction from 7.34 to 5.97. Similarly, FSS scores in the Amantadine group dropped from 53.10 to 28.40, compared to a more modest decrease from 50.38 to 42.59 in the control group.

The treatment proved generally safe and well tolerated, with only one patient discontinuing due to severe side effects. The most common side effect was dry mouth, reported by 26.67% of participants receiving Amantadine. Other side effects included dizziness and headache, but these were generally mild and transient.

While the results suggest Amantadine's potential as an effective treatment for post-COVID-19 fatigue, the study had several limitations. These included its open-label design without placebo control, which could have influenced results through placebo effects. Additionally, the researchers noted that the simple randomization method led to some demographic differences between groups, particularly in age and BMI. The study also lacked long-term follow-up to assess the durability of treatment effects.

The authors conclude by recommending larger, double-blind, placebo-controlled studies to validate their findings and better understand Amantadine's role in treating post-COVID-19 fatigue. This research represents an important step in addressing one of the most common and persistent symptoms affecting COVID-19 survivors.

Other symptom-focused medications, such as those targeting pain, autonomic dysfunction, and cognitive impairment, may be considered for the management of fatigue and associated symptoms in patients with Long COVID. These medications should be used as part of a comprehensive, individualized treatment plan that considers the patient's unique needs and circumstances. Healthcare providers should carefully consider the potential risks and benefits of each medication, considering the patient's comorbidities, contraindications, and medication list. Given the limited evidence for many of these medications in the context of Long COVID, further research is needed to establish their efficacy, safety, and optimal use in this patient population.

16.4 Cannabis as a Potential Treatment

While the mechanisms underlying Long COVID fatigue are not fully understood, some researchers have suggested that cannabis and its active compounds, particularly cannabidiol (CBD) and tetrahydrocannabinol (THC), may have potential therapeutic benefits in managing this symptom [30]. This chapter will explore the current evidence and potential mechanisms of action for cannabis in treating Long COVID fatigue, as well as the limitations and considerations for its use.

16.4.1 Potential Mechanisms of Action

Cannabis contains a variety of active compounds, known as cannabinoids, that interact with the body's endocannabinoid system (ECS) to produce various physiological effects [31]. The ECS is a complex signaling network that plays a role in regulating many processes, including pain, inflammation, mood, and energy metabolism [32]. Some researchers have proposed that dysregulation of the ECS may contribute to the pathophysiology of Long COVID and that targeting this system with cannabinoids may help alleviate fatigue and other symptoms [33].

The two main cannabinoids of interest in the context of Long COVID fatigue are CBD and THC:

a. Cannabidiol (CBD): CBD is a non-psychoactive cannabinoid that has been shown to have anti-inflammatory, analgesic, and anxiolytic properties [34, 35]. CBD may help reduce fatigue in Long COVID by modulating inflammation, reducing pain, and improving sleep quality [36].

b. Tetrahydrocannabinol (THC): THC is the primary psychoactive cannabinoid in cannabis and has been shown to have analgesic, anti-inflammatory, and appetite-stimulating effects [37]. THC may help alleviate fatigue in Long COVID by reducing pain, improving mood, and increasing appetite, which can be impaired in some patients with chronic fatigue [3].

16.4.2 Current Evidence

To date, there is limited direct evidence for the use of cannabis in treating Long COVID fatigue. Most of the available data come from studies on other chronic fatigue conditions, such as chronic fatigue syndrome (CFS) and multiple sclerosis (MS).

Chronic fatigue syndrome: While the current evidence is still limited, the review by Berger et al. [38] suggests a positive effect of cannabis in fibromyalgia and chronic fatigue syndrome. Multiple sclerosis: Several studies have investigated the use of cannabis and cannabinoids for fatigue in MS, with mixed results. A randomized controlled trial by Zajicek et al. [39] found that oral THC was associated with a significant reduction in fatigue compared to placebo in patients with MS. However, a later trial by Zajicek et al. [39] found no significant difference between oral THC and placebo in reducing fatigue. A systematic review by Torres-Moreno et al. [40] concluded that the evidence for cannabis and cannabinoids in treating MS-related fatigue was of low quality and insufficient to make firm conclusions.

Long COVID Fatigue: Two recent papers explore the potential role of cannabinoids, particularly CBD, in treating anxiety disorders related to COVID 19 and post-COVID condition (PCC). The first paper by O'Sullivan et al. [41], published in Cannabis and Cannabinoid Research, discusses how CBD could potentially treat COVID-19-related anxiety disorders. The authors note that CBD has shown anxiolytic properties in clinical trials at doses of 300-800 mg, reducing anxiety in various populations including healthy volunteers and patients with social anxiety disorder. The paper highlights CBD's mechanisms of action through multiple molecular targets, including CB1/CB2 receptors, serotonin 1A receptor, and PPAR-gamma.

The second paper by Cárdenas-Rodríguez et al. [42], published in the International Journal of Molecular Sciences, provides a comprehensive review of cannabis compounds' potential role in managing neuroinflammation in PCC patients. The authors explain that PCC is characterized by persistent symptoms following COVID-19 infection, with neurological manifestations being major complications. The paper details how neuroinflammation and oxidative stress are key factors in PCC pathophysiology, involving microglial activation, BBB disruption, and elevated inflammatory cytokines.

The research suggests that cannabis compounds, including CBD, CBG, and THC, demonstrate significant antioxidative and anti-inflammatory properties that could benefit PCC patients. Clinical trials discussed in the papers show promising results for CBD in treating various neuroinflammatory conditions. For example, one randomized clinical trial showed that a CBD-rich formula improved symptoms in PCC patients, particularly affecting dyspnea, anxiety, and sleep disturbances.

Both papers emphasize the need for further research through well-designed clinical trials to validate these findings. Current ongoing trials are evaluating various cannabis-based treatments for PCC, including studies examining different dosage forms and combinations of cannabinoids. The papers note that while cannabis compounds show promise as therapeutic agents for COVID-related conditions, more evidence is needed to establish optimal treatment protocols and fully understand their mechanisms of action in treating PCC-related neuroinflammation.

These papers represent significant contributions to understanding potential therapeutic approaches for COVID-19 and PCC-related neurological symptoms, while acknowledging the current limitations and need for additional research in this emerging field.

The two papers provide important insights into how cannabinoids might address Long COVID fatigue through multiple mechanisms. O'Sullivan et al. (2021) and Cárdenas-Rodríguez et al. (2024) both connect neuroinflammation and oxidative stress to the persistent symptoms of Long COVID, including fatigue.

The association between cannabinoids and Long COVID fatigue can be understood through several key pathways discussed in these papers:

Neuroinflammatory Mechanisms: Cárdenas-Rodríguez et al. [41] explain that Long COVID fatigue may be linked to sustained neuroinflammation, where inflammatory cytokines like IL-6, IL-1β, and TNF-α remain elevated. The papers demonstrate that CBD and other cannabinoids can reduce these inflammatory markers, potentially alleviating fatigue symptoms. This is particularly relevant as the authors note that chronic neuroinflammation can lead to persistent fatigue through disruption of normal neural functioning.

Oxidative Stress: Both papers highlight how oxidative stress contributes to Long COVID symptoms. Cárdenas-Rodríguez et al. [41] specifically detail how SARS-CoV-2 infection leads to excessive production of reactive oxygen species (ROS) and reactive nitrogen species (RNS), which can cause cellular damage and contribute to fatigue. The antioxidant properties of cannabinoids, particularly CBD and CBG, could help combat this oxidative damage.

Mitochondrial Function: The papers suggest that Long COVID fatigue may be related to mitochondrial dysfunction. Cárdenas-Rodríguez et al. [41] note that cannabis compounds might help restore normal mitochondrial function through their anti-inflammatory and antioxidant effects, potentially improving energy metabolism and reducing fatigue.

Clinical Evidence: The papers discuss several clinical trials investigating cannabinoids for Long COVID symptoms. One notable trial mentioned in Cárdenas-Rodríguez et al. [41] found that a CBD-rich formula improved various symptoms in Long COVID patients, including fatigue. The authors cite a study where 65% of patients with Long COVID fatigue responded positively to treatment protocols that included anti-inflammatory approaches.

Sleep Regulation: O'Sullivan et al. [42] discuss how CBD can improve sleep quality, which is often disrupted in Long COVID patients. Better sleep regulation through cannabinoid therapy could indirectly help manage fatigue symptoms.

However, both papers emphasize that while these mechanisms are promising, more research is needed to fully understand how cannabinoids can best be used to treat Long COVID fatigue. They call for larger, well-designed clinical trials specifically focused on fatigue outcomes. The papers also note the importance of considering individual patient factors and potential drug interactions when using cannabinoids for Long COVID fatigue treatment.

These findings suggest that cannabinoids could offer a multi-targeted approach to treat Long COVID fatigue by addressing both the underlying inflammation and oxidative stress while potentially improving sleep and energy metabolism. However, the authors emphasize that this remains an emerging area of research requiring further investigation to establish optimal treatment protocols.

The papers encourage healthcare providers to consider the potential role of cannabinoids as part of a comprehensive treatment approach for Long COVID fatigue, while acknowledging the need for more evidence from ongoing clinical trials. This research contributes to our understanding of how cannabinoid-based therapies might help address one of the most challenging aspects of Long COVID syndrome.

The two papers provide important insights into how cannabinoids might address Long COVID fatigue through multiple mechanisms. O'Sullivan et al. [41] and Cárdenas-Rodríguez et al. [42] both connect neuroinflammation and oxidative stress to the persistent symptoms of Long COVID, including fatigue.

The association between cannabinoids and Long COVID fatigue can be understood through several key pathways discussed in these papers:

Neuroinflammatory Mechanisms: Cárdenas-Rodríguez et al. [42] explain that Long COVID fatigue may be linked to sustained neuroinflammation, where inflammatory cytokines like IL-6, IL-1β, and TNF-α remain elevated, as demonstrated by Gebo et al. [43] in their randomized clinical trial. The papers demonstrate that CBD and other cannabinoids can reduce these inflammatory markers, as shown by Barichello et al. [44] in their study of TNF-α reduction in the frontal cortex, potentially alleviating fatigue symptoms. This is particularly relevant as Elliott et al. [45] noted that chronic neuroinflammation can lead to persistent fatigue through disruption of normal neural functioning.

Oxidative Stress: Both papers highlight how oxidative stress contributes to Long COVID symptoms. Cárdenas-Rodríguez et al. [42] specifically detail how SARS-CoV-2 infection leads to excessive production of reactive oxygen species (ROS) and reactive nitrogen species (RNS), supported by research from Gain et al. [46]. The antioxidant properties of cannabinoids, particularly CBD and CBG, could help combat this oxidative damage, as demonstrated by Perez et al. [47] in their in vitro studies.

Mitochondrial Function: The papers suggest that Long COVID fatigue may be related to mitochondrial dysfunction. Georgieva et al. [48] showed how

COVID-19 affects mitochondrial and endothelial dysfunction. Cárdenas-Rodríguez et al. [42] note that cannabis compounds might help restore normal mitochondrial function through their anti-inflammatory and antioxidant effects, potentially improving energy metabolism and reducing fatigue.

Clinical Evidence: Young et al. [49] conducted a single-blind, randomized, placebo-controlled study showing Formula C™ (CBD-rich) improved various symptoms in Long COVID patients. Binetti et al. [50] found that 65% of patients with Long COVID fatigue responded positively to treatment protocols that included anti-inflammatory approaches.

Sleep Regulation: O'Sullivan et al. [41] cite studies by Shannon et al. [51] and Carlini and Cunha [52] showing how CBD can improve sleep quality at doses of 25-175 mg and 160 mg respectively, which is often disrupted in Long COVID patients.

The papers document several ongoing clinical trials (NCT04997395, NCT05467904, NCT04777981) investigating cannabis-based treatments for PCC. These trials should provide more definitive evidence regarding the effectiveness of cannabinoid-based treatments for fatigue and other PCC symptoms.

Both papers emphasize that while these mechanisms are promising, more research is needed to fully understand how cannabinoids can best be used to treat Long COVID fatigue. Cinar et al. [53] specifically call for investigations into dual inhibition of CB1 receptors and iNOS as a potential approach to managing Long COVID-19 symptoms. Davis et al. [54] emphasize the importance of considering individual patient factors and potential drug interactions, particularly given CBD's effects on liver enzymes as documented by Chesney et al. [55] in their systematic review of CBD's adverse effects.

These findings suggest that cannabinoids could offer a multi-targeted approach to treating Long COVID fatigue. Wang et al. [56] demonstrated CBD's ability to alleviate neuroinflammation through targeting FKBP5, while Sasso et al. [57] identified TRPM3 dysfunction as a potential therapeutic target in PCC patients. The anti-inflammatory properties of cannabinoids are further supported by van Breemen et al. [58], who showed cannabinoid acids could block cellular entry of SARS-CoV-2.

Alternative approaches are also being investigated. Di Stadio et al. [59] evaluated PEALUT (palmitoylethanolamide with luteolin) for PCC symptoms, while Cash and Kaufman [60] demonstrated oxaloacetate's effectiveness in reducing fatigue in 46.8% of PCC patients. Ghali et al. [61] and Parker et al. [62] highlighted the importance of pacing strategies in managing post-exertional symptom exacerbation, suggesting that any cannabinoid treatment might be most effective as part of a comprehensive treatment approach.

The papers note several promising ongoing clinical trials. Young et al. [63] are investigating Formula C™, while other researchers are examining different cannabis formulations through trials NCT04997395 (feasibility study of CBD for Long COVID), NCT05467904 (trial of hemp flower formulation), and NCT04777981 (study of CBDRA60-CBD with red algae). As emphasized by Munblit et al. [64] in their international Delphi consensus study, standardized outcome measures for PCC trials are crucial for advancing our understanding of treatment efficacy.

This research contributes significantly to our understanding of how cannabinoid-based therapies might help address Long COVID fatigue, supporting the observations of Al-Hakeim et al. [65] regarding the role of oxidative damage and inflammation in chronic post-viral fatigue symptoms. The work of Nguyen et al. [66] on CBD's ability to inhibit SARS-CoV-2 replication through modulation of host stress responses further suggests potential preventive applications alongside therapeutic ones.

16.4.3 Limitations and Considerations

While cannabis may have potential therapeutic benefits for Long COVID fatigue, there are several limitations and considerations that should be considered: (a). Limited evidence: As mentioned above, there is currently limited direct evidence for the use of cannabis in treating Long COVID fatigue. More research, including randomized controlled trials, is needed to establish its efficacy and safety in this specific population. (b). Potential side effects: Cannabis and cannabinoids can cause various side effects, including dizziness, dry mouth, nausea, and impaired cognition and coordination [67]. These side effects may be particularly problematic for patients with Long COVID who are already experiencing cognitive impairment or other neurological symptoms. (c). Drug interactions: Cannabinoids can interact with various medications, including anticoagulants, antidepressants, and anti-seizure drugs, potentially altering their effectiveness or increasing the risk of side effects [68]. Patients with Long COVID who are taking other medications should discuss the potential for drug interactions with their healthcare provider before using cannabis. (d). Legal and accessibility issues: The legal status of cannabis varies widely across countries and jurisdictions, which can impact its accessibility and use for medical purposes [69]. In areas where cannabis is illegal or heavily restricted, patients with Long COVID may have difficulty accessing it as a potential treatment option. (e). Stigma and social acceptance: Despite growing acceptance of medical cannabis, there is still significant stigma and social disapproval surrounding its use [69]. Patients with Long COVID who use cannabis to manage their symptoms may face judgment or discrimination from others, which can impact their social and occupational functioning.

Cannabis and its active compounds, particularly CBD and THC, may have potential therapeutic benefits for managing fatigue in patients with Long COVID. The proposed mechanisms of action include modulating inflammation, reducing pain, improving sleep quality, and regulating energy metabolism through the endocannabinoid system. However, the current evidence for the use of cannabis in treating Long COVID fatigue is limited, and more research is needed to establish its efficacy and safety in this specific population. Patients and healthcare providers should also consider the potential side effects, drug interactions, legal and accessibility issues, and social stigma associated with cannabis use. As with any potential treatment option, the decision to use cannabis for Long COVID fatigue should be made on a case-by-case basis, weighing the potential benefits and risks, and in consultation with a qualified healthcare provider.

16.5 Noninvasive Brain Stimulation

Some researchers have suggested that noninvasive brain stimulation (NIBS) techniques, such as transcranial direct current stimulation (tDCS), may have potential therapeutic benefits in managing this symptom [70]. This chapter will explore the current evidence and potential mechanisms of action for NIBS in treating Long COVID fatigue, as well as the limitations and considerations for its use.

1. Noninvasive Brain Stimulation Techniques: NIBS techniques, such as tDCS, involve the application of magnetic fields or weak electrical currents to specific areas of the brain to modulate neural activity and plasticity [71]. These techniques have been investigated as potential treatments for various neurological and psychiatric conditions, including depression, chronic pain, and fatigue [72]. tDCS involves the application of weak electrical currents (typically 1–2 mA) to the brain via electrodes placed on the scalp [73]. Anodal tDCS is thought to increase neural excitability, while cathodal tDCS is thought to decrease it [74]. tDCS has been investigated as a potential treatment for fatigue in various conditions, such as multiple sclerosis (MS) and chronic fatigue syndrome (CFS) [75].
2. Potential Mechanisms of Action: The exact mechanisms by which NIBS techniques may alleviate fatigue in Long COVID are not fully understood, but several potential pathways have been proposed: a. Modulation of cortical excitability: Fatigue has been associated with alterations in cortical excitability, particularly in the prefrontal cortex and motor cortex [76]. tDCS may help normalize cortical excitability and improve fatigue by targeting these regions [75]. b. Regulation of neurotransmitter systems: Fatigue has been linked to dysregulation of neurotransmitter systems, including dopamine, serotonin, and norepinephrine [77]. tDCS may modulate the activity of these neurotransmitter systems and improve fatigue by targeting relevant brain regions, such as the dorsolateral prefrontal cortex (DLPFC) [78]. c. Reduction of inflammation: Chronic inflammation has been implicated in the pathophysiology of Long COVID and may contribute to fatigue [79]. Linnhoff et al. [70] have suggested that tDCS may have anti-inflammatory effects and could potentially reduce fatigue by modulating immune function.
3. Current Evidence: Klírová et al. [80] conducted a double-blind, randomized, sham-controlled study evaluating tDCS for treating neuropsychiatric symptoms in Long COVID patients. The study included 33 patients (16 active, 17 sham) who received 20 tDCS sessions over 4 weeks targeting the prefrontal cortex. The primary outcome was a change in fatigue measured by the Fatigue Impact Scale (FIS). Interestingly, while both groups showed improvement, the sham group actually had a greater decrease in FIS scores, though the difference between

groups was not statistically significant. The study found no significant differences between active and sham tDCS for anxiety, depression, quality of life, or cognitive performance. The authors note several limitations that may have affected results, including small sample size, baseline FIS score differences between groups, and chosen stimulation parameters. They conclude that either the study design had limitations or tDCS may be insufficient for treating these symptoms. Santana et al. [81] evaluated high-definition tDCS (HD-tDCS) combined with rehabilitation for treating fatigue in Long COVID patients. This larger study included 70 patients randomized to receive either active or sham HD-tDCS targeting the left primary motor cortex, combined with rehabilitation exercises, for 10 sessions over 5 weeks. In contrast to the first study, this trial found significant benefits from active treatment. The active HD-tDCS group showed greater reduction in fatigue compared to sham (22.11 vs 10.34 point reduction on the Modified Fatigue Impact Scale). The benefits were particularly notable for cognitive and psychosocial fatigue domains. The active group also showed greater improvements in anxiety and quality of life, though pain levels were similar between groups. The treatment was well tolerated with only minor side effects like skin redness. The contrasting results between these studies highlight several important factors:

(a) Different stimulation targets (prefrontal cortex vs motor cortex)
(b) Different protocols (HD-tDCS vs conventional tDCS)
(c) Integration with rehabilitation in the second study
(d) Larger sample size in the second study
(e) Different outcome measures and assessment tools

Both studies contribute valuable data to understanding the potential role of brain stimulation in treating Long COVID symptoms, while also demonstrating the importance of optimization of treatment parameters and protocols. The positive results from the HD-tDCS study suggest this may be a promising therapeutic approach when properly implemented and combined with rehabilitation.

16.6 A Forward-Looking Perspective on AI-Guided tDCS in Long COVID Management

The integration of artificial intelligence with transcranial direct current stimulation represents a promising frontier in Long COVID treatment. As Rudroff, Rainio, and Klén [82] comprehensively outline in their perspective review, AI could fundamentally transform how we approach tDCS optimization and delivery. This synergy between AI and neuromodulation offers unprecedented opportunities for personalizing treatment approaches while addressing the complex, heterogeneous nature of Long COVID symptoms.

At the core of this advancement lies the potential for precise treatment personalization. AI algorithms can analyze vast arrays of patient data—from neuroimaging results to clinical symptoms and real-time physiological measurements—to determine optimal stimulation parameters for each individual. This capability is particularly crucial given the marked variability in Long COVID presentations and the known heterogeneity in tDCS response patterns [81]. The technology could enable dynamic adjustment of stimulation parameters based on ongoing patient responses, potentially improving treatment efficacy.

Clinical trial design stands to benefit significantly from AI integration. Machine learning approaches can facilitate more sophisticated patient stratification, identifying subgroups most likely to benefit from specific tDCS protocols. This could lead to more efficient and targeted trials, potentially accelerating the development of evidence-based treatment protocols. AI-driven adaptive trial designs could optimize protocol parameters more rapidly than traditional approaches, while simultaneously developing more sensitive outcome measures [82].

However, significant challenges must be addressed for successful implementation. Data quality and quantity requirements present a particular hurdle, especially given the novelty of Long COVID as a condition. Privacy concerns and ethical considerations surrounding AI in healthcare must be carefully navigated, particularly regarding data protection and algorithmic bias. The authors emphasize the critical importance of maintaining equitable access to treatment while implementing these advanced technologies.

Looking toward the future, the authors project that by 2050, we might see fully integrated diagnostics combining genomic, proteomic, and metabolic neural imaging data to guide tDCS protocols. This could enable precise treatments to disease endotypes based on underlying biology. The development of closed-loop stimulation systems, incorporating real-time fMRI with integrated EEG inputs, could provide unprecedented control over treatment delivery and optimization.

As we move forward with this promising approach, maintaining a balance between innovation and practical clinical utility will be crucial. Success will require careful attention to implementation challenges, including healthcare system integration, clinician training, and cost considerations. Furthermore, as both manuscripts emphasize, the development of standardized protocols for data collection and analysis across multiple sites will be essential for building the robust evidence base needed to support widespread adoption.

Summary

This chapter examines the current landscape of treatments for Long COVID-related fatigue, presenting a detailed analysis of both established and emerging therapeutic approaches. The chapter begins by exploring the role of stimulant medications, particularly methylphenidate and modafinil, in managing severe fatigue symptoms. While these medications show promise in small-scale studies, their use requires careful consideration of individual patient factors, potential side effects, and the current limitations of clinical evidence.

The discussion then broadens to encompass a range of symptom-focused medications, including treatments for pain, autonomic dysfunction, and cognitive impairment. Of particular interest is the emerging evidence supporting low-dose naltrexone (LDN) combined with NAD+ supplementation, which has shown encouraging results in recent clinical trials for managing persistent fatigue symptoms. The chapter provides a thorough examination of autonomic dysfunction treatments, including beta-blockers and medications for orthostatic intolerance, reflecting the complex nature of Long COVID symptoms.

A significant portion of the chapter is dedicated to exploring the potential therapeutic role of cannabis and cannabinoids in treating Long COVID fatigue. The chapter delves into the mechanisms by which CBD and THC might address underlying inflammation and oxidative stress, supported by recent research indicating multiple pathways through which cannabinoids could potentially alleviate fatigue symptoms. A balanced view of both the promising aspects and limitations of cannabis-based treatments are presented, including important considerations regarding legal status, accessibility, and social stigma.

The chapter concludes with an innovative examination of noninvasive brain stimulation techniques, particularly transcranial direct current stimulation (tDCS). This section presents contrasting findings from recent clinical trials and explores the exciting potential of AI-guided neuromodulation in future treatment protocols. The review maintains a critical perspective throughout, acknowledging the preliminary nature of much of the available evidence while highlighting the need for larger, well-designed clinical trials to establish definitive treatment guidelines.

Throughout the chapter, there is a consistent emphasis on the importance of individualized treatment approaches and the need to carefully weigh potential benefits against risks for each intervention. The review bridges the gap between current clinical practice and emerging research, providing valuable insights for healthcare providers managing patients with Long COVID fatigue.

References

1. Greenhalgh T, Knight M, A'Court C, Buxton M, Husain L. Management of post-acute covid-19 in primary care. BMJ. 2020;370:m3026.
2. Yang TT, Wang L, Deng XY, Yu G. Pharmacological treatments for fatigue in patients with multiple sclerosis: A systematic review and meta-analysis. J Neurol Sci. 2017;380:256–61.
3. Blockmans D, Persoons P. Long-term methylphenidate intake in chronic fatigue syndrome. Acta Clin Belg. 2016;71:407. https://doi.org/10.1080/17843286.2016.1200816.
4. Zifko UA, Rupp M, Schwarz S, Zipko HT, Maida EM. Modafinil in treatment of fatigue in multiple sclerosis: Results of an open-label study. J Neurol. 2002;249(8):983–7.
5. Calabrese JR, Durgam S, Satlin A, Vanover KE, Davis RE, Chen R, et al. Efficacy and safety of lumateperone for major depressive episodes associated with bipolar I or bipolar II disorder: a phase 3 randomized placebo-controlled trial. Am J Psychiatry. 2021;178(12):1098–106.
6. Bhattacharjee D, Surakshitha Poornima HK, Chakraborty A. Methylphenidate in COVID-19 related brain fog: A case series. Indian J Psychol Med. 2024;46(1):02537176231222572.

7. Miskowiak KW, Johnsen S, Sattler SM, Nielsen S, Kunalan K, Rungby J, et al. Cognitive impairments four months after COVID-19 hospital discharge: Pattern, severity and association with illness variables. Eur Neuropsychopharmacol. 2021;46:39–48.
8. Nalbandian A, Sehgal K, Gupta A, Madhavan MV, McGroder C, Stevens JS, et al. Post-acute COVID-19 syndrome. Nat Med. 2021;27(4):601–15.
9. Crook H, Raza S, Nowell J, Young M, Edison P. Long COVID—mechanisms, risk factors, and management. BMJ. 2021;374:n1648.
10. Karaarslan F, Demircioğlu Güneri F, Kardeş S. Postdischarge rheumatic and musculoskeletal symptoms following hospitalization for COVID-19: prospective follow-up by phone interviews. Rheumatol Int. 2021;41(7):1263–71.
11. Bunchorntavakul C, Reddy KR. Acetaminophen (APAP or N-acetyl-p-aminophenol) and acute liver failure. Clin Liver Dis. 2018;22(2):325–46.
12. Bates D, Schultheis BC, Hanes MC, Jolly SM, Chakravarthy KV, Deer TR, et al. A comprehensive algorithm for management of neuropathic pain. Pain Med. 2019;20(Suppl 1):S2–S12.
13. Wijeratne T, Crewther S. COVID-19 and long-term neurological problems: Challenges ahead with Post-COVID-19 Neurological Syndrome. Aust J Gen Pract. 2021;50:50. https://doi.org/10.31128/AJGP-COVID-43.
14. Toljan K, Vrooman B. Low-dose naltrexone (LDN)—review of therapeutic utilization. Med Sci (Basel). 2018;6(4):82.
15. Younger J, Parkitny L, McLain D. The use of low-dose naltrexone (LDN) as a novel anti-inflammatory treatment for chronic pain. Clin Rheumatol. 2014;33(4):451–9.
16. Brown N, Panksepp J. Low-dose naltrexone for disease prevention and quality of life. Med Hypotheses. 2009;72(3):333–7.
17. Isman A, Nyquist A, Strecker B, Harinath G, Lee V, Zhang X, et al. Low-dose naltrexone and NAD+ for the treatment of patients with persistent fatigue symptoms after COVID-19. Brain Behav Immun Health. 2024;36:100733.
18. Yong SJ. Persistent brainstem dysfunction in long-COVID: A hypothesis. ACS Chem Neurosci. 2021;12(4):573–80.
19. Bisaccia G, Ricci F, Recce V, Serio A, Iannetti G, Chahal AA, et al. Post-acute sequelae of COVID-19 and cardiovascular autonomic dysfunction: what do we know? J Cardiovasc Dev Dis. 2021;8(11):156.
20. Bryarly M, Phillips LT, Fu Q, Vernino S, Levine BD. Postural Orthostatic Tachycardia Syndrome: JACC Focus Seminar. J Am Coll Cardiol. 2019;73(10):1207–28.
21. Vernino S, Bourne KM, Stiles LE, Grubb BP, Fedorowski A, Stewart JM, et al. Postural orthostatic tachycardia syndrome (POTS): State of the science and clinical care from a 2019 National Institutes of Health Expert Consensus Meeting - Part 1. Auton Neurosci. 2021;235:102828.
22. Larsen NW, Stiles LE, Miglis MG. Preparing for the long-haul: Autonomic complications of COVID-19. Auton Neurosci. 2021;235:102841.
23. Raj SR, Arnold AC, Barboi A, Claydon VE, Limberg JK, Lucci VM, et al. Long-COVID postural tachycardia syndrome: an American Autonomic Society statement. Clin Auton Res. 2020;31(3):365–8.
24. Graham EL, Clark JR, Orban ZS, Lim PH, Szymanski AL, Taylor C, et al. Persistent neurologic symptoms and cognitive dysfunction in non-hospitalized Covid-19 "long haulers". Ann Clin Transl Neurol. 2021;8(5):1073–85.
25. Birks JS, Harvey RJ. Donepezil for dementia due to Alzheimer's disease. Cochrane Database Syst Rev. 2018;6:CD001190.
26. Hampel H, Mesulam MM, Cuello AC, Farlow MR, Giacobini E, Grossberg GT, et al. The cholinergic system in the pathophysiology and treatment of Alzheimer's disease. Brain. 2018;141(7):1917–33.
27. Almeria M, Cejudo JC, Sotoca J, Deus J, Krupinski J. Cognitive profile following COVID-19 infection: Clinical predictors leading to neuropsychological impairment. Brain Behav Immun Health. 2020;9:100163.
28. Maldonado R, Baños JE, Cabañero D. The endocannabinoid system and neuropathic pain. Pain. 2016;157(Suppl 1):S23–32.

29. Harandi AA, Pakdaman H, Medghalchi A, Kimia N, Kazemian A, Siavoshi F, Barough SS, Esfandani A, Hosseini MH, Sobhanian SA. A randomized open-label clinical trial on the effect of Amantadine on post Covid 19 fatigue. Sci Rep. 2024;14(1):1343. https://doi.org/10.1038/s41598-024-51904-z. PMID: 38228731; PMCID: PMC10792128.
30. Zou S, Kumar U. Cannabinoid receptors and the endocannabinoid system: signaling and function in the central nervous system. Int J Mol Sci. 2018;19(3):833.
31. Aizpurua-Olaizola O, Elezgarai I, Rico-Barrio I, Zarandona I, Etxebarria N, Usobiaga A. Targeting the endocannabinoid system: future therapeutic strategies. Drug Discov Today. 2017;22(1):105–10.
32. Scott C, Hall S, Zhou J, Lehmann C. Cannabinoids and the endocannabinoid system in early SARS-CoV-2 infection and long COVID-19-A scoping review. J Clin Med. 2023;13(1):227.
33. Blessing EM, Steenkamp MM, Manzanares J, Marmar CR. Cannabidiol as a potential treatment for anxiety disorders. Neurotherapeutics. 2015;12(4):825–36.
34. Hammell DC, Zhang LP, Ma F, Abshire SM, McIlwrath SL, Stinchcomb AL, et al. Transdermal cannabidiol reduces inflammation and pain-related behaviours in a rat model of arthritis. Eur J Pain. 2016;20(6):936–48.
35. Shannon S, Lewis N, Lee H, Hughes S. Cannabidiol in anxiety and sleep: a large case series. Perm J. 2019;23:18–041.
36. Abrams DI. The therapeutic effects of Cannabis and cannabinoids: an update from the National Academies of Sciences, Engineering and Medicine report. Eur J Intern Med. 2018;49:7–11.
37. Suraev AS, Marshall NS, Vandrey R, McCartney D, Benson MJ, McGregor IS, et al. Cannabinoid therapies in the management of sleep disorders: a systematic review of preclinical and clinical studies. Sleep Med Rev. 2020;53:101339.
38. Berger AA, Keefe J, Winnick A, Gilbert E, Eskander JP, Yazdi C, et al. Cannabis and cannabidiol (CBD) for the treatment of fibromyalgia. Best Pract Res Clin Anaesthesiol. 2020;34(3):617–31.
39. Zajicek J, Ball S, Wright D, Vickery J, Nunn A, Miller D, et al. Effect of dronabinol on progression in progressive multiple sclerosis (CUPID): a randomized, placebo-controlled trial. Lancet Neurol. 2013;12(9):857–65.
40. Torres-Moreno MC, Papaseit E, Torrens M, Farré M. Assessment of efficacy and tolerability of medicinal cannabinoids in patients with multiple sclerosis: a systematic review and meta-analysis. JAMA Netw Open. 2018;1(6):e183485.
41. O'Sullivan SE, Stevenson CW, Laviolette SR. Could cannabidiol be a treatment for coronavirus disease-19-related anxiety disorders? Cannabis Cannabinoid Res. 2021;6(1):7–18.
42. Cárdenas-Rodríguez N, Ignacio-Mejía I, Correa-Basurto J, Carrasco-Vargas H, Vargas-Hernández MA, Albores-Méndez EM, et al. Possible role of cannabis in the management of neuroinflammation in patients with post-COVID condition. Int J Mol Sci. 2024;25(7):3805.
43. Gebo KA, Heath SL, Fukuta Y, Zhu X, Baksh S, Abraham AG, et al. Early antibody treatment, inflammation, and risk of post-COVID conditions. MBio. 2023;14:e0061823.
44. Barichello T, Ceretta RA, Generoso JS, Moreira AP, Simões LR, Comim CM, et al. Cannabidiol reduces host immune response and prevents cognitive impairments in Wistar rats submitted to pneumococcal meningitis. Eur J Pharmacol. 2012;697:158–64.
45. Elliott DM, Singh N, Nagarkatti M, Nagarkatti PS. Cannabidiol attenuates experimental autoimmune encephalomyelitis model of multiple sclerosis through induction of myeloid-derived suppressor cells. Front Immunol. 2018;9:1782.
46. Gain C, Song S, Angtuaco T, Satta S, Kelesidis T. The role of oxidative stress in the pathogenesis of infections with coronaviruses. Front Microbiol. 2023;13:1111930.
47. Perez E, Fernandez JR, Fitzgerald C, Rouzard K, Tamura M, Savile C. In vitro and clinical evaluation of cannabigerol (CBG) produced via yeast biosynthesis: A Cannabinoid with a broad range of anti-inflammatory and skin health-boosting properties. Molecules. 2022;27:491.
48. Georgieva E, Ananiev J, Yovchev Y, Arabadzhiev G, Abrashev H, Abrasheva D, et al. COVID-19 complications: oxidative stress, inflammation, and mitochondrial and endothelial dysfunction. Int J Mol Sci. 2023;24:14876.

49. Young TP, Erickson JS, Hattan SL, Guzy S, Hershkowitz F, Steward MD. A single-blind, randomized, placebo controlled study to evaluate the benefits and safety of endourage targeted Wellness Formula C sublingual+ Drops in people with post-acute coronavirus disease 2019 syndrome. Cannabis Cannabinoid Res. 2022;9:282–92.
50. Binetti J, Real M, Renzulli M, Bertran L, Riesco D, Perpiñan C, et al. Clinical and biomarker profile responses to rehabilitation treatment in patients with long COVID characterized by chronic fatigue. Viruses. 2023;15:1452.
51. Shannon S, Opila-Lehman J. Effectiveness of cannabidiol oil for pediatric anxiety and insomnia as part of posttraumatic stress disorder: a case report. Perm J. 2016;20:16–005.
52. Carlini EA, Cunha JM. Hypnotic and antiepileptic effects of cannabidiol. J Clin Pharmacol. 1981;21:417S–27S.
53. Cinar R, Iyer MR, Kunos G. Dual inhibition of CB1 receptors and iNOS, as a potential novel approach to the pharmacological management of acute and long COVID-19. Br J Pharmacol. 2022;179(10):2121–7.
54. Davis HE, McCorkell L, Vogel JM, Topol EJ. Long COVID: Major findings, mechanisms and recommendations. Nat Rev Microbiol. 2023;21:133–46.
55. Chesney E, Oliver D, Green A, Sovi S, Wilson J, Englund A, et al. Adverse effects of cannabidiol: a systematic review and meta-analysis of randomized clinical trials. Neuropsychopharmacology. 2020;45(11):1799–806.
56. Wang X, Lin C, Jin S, Wang Y, Peng Y, Wang X. Cannabidiol alleviates neuroinflammation and attenuates neuropathic pain via targeting FKBP5. Brain Behav Immun. 2023;111:365–75.
57. Sasso EM, Muraki K, Eaton-Fitch N, Smith P, Lesslar OL, Deed G, et al. Transient receptor potential melastatin 3 dysfunction in post COVID-19 condition and myalgic encephalomyelitis/chronic fatigue syndrome patients. Mol Med. 2022;28:98.
58. van Breemen RB, Muchiri RN, Bates TA, Weinstein JB, Leier HC, Farley S, et al. Cannabinoids block cellular entry of SARS-CoV-2 and the emerging variants. J Nat Prod. 2022;85:176–84.
59. Di Stadio A, Gallina S, Cocuzza S, De Luca P, Ingrassia A, Oliva S, et al. Treatment of COVID-19 olfactory dysfunction with olfactory training, palmitoylethanolamide with luteolin, or combined therapy: A blinded controlled multicenter randomized trial. Eur Arch Otorrinolaringol. 2023;280:4949–56.
60. Cash A, Kaufman DL. Oxaloacetate Treatment For Mental And Physical Fatigue In Myalgic Encephalomyelitis/Chronic Fatigue Syndrome (ME/CFS) and Long-COVID fatigue patients: A non-randomized controlled clinical trial. J Transl Med. 2022;20:295.
61. Ghali A, Lacombe V, Ravaiau C, Delattre E, Ghali M, Urbanski G, et al. The relevance of pacing strategies in managing symptoms of post-COVID-19 syndrome. J Transl Med. 2023;21:375.
62. Parker M, Sawant HB, Flannery T, Tarrant R, Shardha J, Bannister R, et al. Effect of using a structured pacing protocol on post-exertional symptom exacerbation and health status in a longitudinal cohort with the post-COVID-19 syndrome. J Med Virol. 2023;95:e28373.
63. Munblit D, Nicholson T, Akrami A, Apfelbacher C, Chen J, De Groote W, et al. A core outcome set for post-COVID-19 condition in adults for use in clinical practice and research: An international Delphi consensus study. Lancet Respir Med. 2022;10:715–24.
64. Al-Hakeim HK, Al-Rubaye HT, Al-Hadrawi DS, Almulla AF, Maes M. Chronic post-viral fatigue and affective symptoms of prolonged COVID are associated with oxidative damage, decreased antioxidant defenses, and inflammation: a proof of concept and mechanism study. Mol Psychiatry. 2023;28:564–78.
65. Nguyen LC, Yang D, Nicolaescu V, Best TJ, Gula H, Saxena D, et al. Cannabidiol inhibits SARS-CoV-2 replication through induction of the host ER stress and innate immune responses. Sci Adv. 2022;8:eabi6110.
66. Whiting PF, Wolff RF, Deshpande S, Di Nisio M, Duffy S, Hernandez AV, et al. Cannabinoids for medical use: a systematic review and meta-analysis. JAMA. 2015;313(24):2456–73.
67. Alsherbiny MA, Li CG. Medicinal cannabis—potential drug interactions. Medicines. 2019;6:3.
68. Abuhasira R, Shbiro L, Landschaft Y. Medical use of cannabis and cannabinoids containing products - Regulations in Europe and North America. Eur J Intern Med. 2018;49:2–6.

69. Bottorff JL, Bissell LJ, Balneaves LG, Robinson CA, Gibson H, Fisher K, et al. Perceptions of cannabis as a stigmatized medicine: a qualitative descriptive study. Harm Reduct J. 2013;10:2.
70. Linnhoff S, Koehler L, Haghikia A, Zaehle T. The therapeutic potential of non-invasive brain stimulation for the treatment of Long-COVID-related cognitive fatigue. Front Immunol. 2023;13:935614.
71. Bikson M, Grossman P, Thomas C, Zannou AL, Jiang J, Adnan T, et al. Safety of transcranial direct current stimulation: Evidence based update 2016. Brain Stimul. 2016;9(5):641–61.
72. Lefaucheur JP, Chalah MA, Mhalla A, Palm U, Ayache SS, Mylius V. The treatment of fatigue by non-invasive brain stimulation. Neurophysiol Clin. 2017;47(2):173–84.
73. Nitsche MA, Paulus W. Excitability changes induced in the human motor cortex by weak transcranial direct current stimulation. J Physiol. 2000;527(3):633–9.
74. Nitsche MA, Cohen LG, Wassermann EM, Priori A, Lang N, Antal A, et al. Transcranial direct current stimulation: State of the art 2008. Brain Stimul. 2008;1(3):206–23.
75. Ayache SS, Chalah MA, Kümpfel T, Padberg F, Lefaucheur JP, Palm U. Multiple sclerosis fatigue, its neural correlates, and its modulation with tDCS. Fortschr Neurol Psychiatr. 2017;85(5):260–9.
76. Kluger BM, Krupp LB, Enoka RM. Fatigue and fatigability in neurologic illnesses: proposal for a unified taxonomy. Neurology. 2013;80(4):409–16.
77. Dobryakova E, Genova HM, DeLuca J, Wylie GR. The dopamine imbalance hypothesis of fatigue in multiple sclerosis and other neurological disorders. Front Neurol. 2015;6:52.
78. Hone-Blanchet A, Edden RA, Fecteau S. Online effects of transcranial direct current stimulation in real time on human prefrontal and striatal metabolites. Biol Psychiatry. 2016;80(6):432–8.
79. Noviello M, De Lorenzo R, Chimienti R, Maugeri N, De Lalla C, Siracusano G, et al. The longitudinal characterization of immune responses in COVID-19 patients reveals novel prognostic signatures for disease severity, patients' survival and long COVID. Front Immunol. 2024;15:1381091.
80. Klírová M, Adamová A, Biačková N, Laskov O, Renková V, Stuchlíková Z, et al. Transcranial direct current stimulation (tDCS) in the treatment of neuropsychiatric symptoms of long COVID. Sci Rep. 2024;14:2193.
81. Santana K, França E, Sato J, Silva A, Queiroz M, de Farias J, et al. Non-invasive brain stimulation for fatigue in post-acute sequelae of SARS-CoV-2 (PASC). Brain Stimul. 2023;16(1):100–7.
82. Rudroff T, Rainio O, Klén R. Leveraging artificial intelligence to optimize transcranial direct current stimulation for long COVID management: a forward-Looking perspective. Brain Sci. 2024;14:Article 831.

Revealing the Complexity of Long COVID Fatigue: Challenges and Promises of Artificial Intelligence

17

17.1 The Multifaceted Challenge of Long COVID Fatigue Research: A Deep Dive

The complexity of Long COVID fatigue research emerges from an intricate interplay of biological, clinical, and measurement factors that challenge our traditional research paradigms. At the core of this challenge lies the multifaceted nature of fatigue that manifests in unprecedented ways. As Rudroff et al. [1–4] reveal in his Long COVID research, fatigue presentations exhibit not just varying severity but fundamentally different trajectories. Some patients show pronounced cognitive fatigue with distinct metabolic patterns in their basal ganglia, while others present primarily with physical fatigue but maintain normal brain metabolism. This heterogeneity extends beyond simple symptom variation to encompass fundamental differences in underlying fatigue mechanisms.

The integration of multimodal biomarkers presents particularly fascinating challenges in understanding fatigue mechanisms. Consider the findings from FDG-PET studies: Guedj et al. [5] demonstrated significant hypometabolism in brain regions associated with energy regulation and motivation, while Rudroff et al. [6] revealed that metabolic heterogeneity might actually serve as a protective mechanism against fatigue development. These seemingly contradictory findings highlight how different biomarkers can tell disparate parts of the same story. When incorporating additional measures—inflammatory markers associated with fatigue severity, autonomic dysfunction patterns, and structural brain changes seen on MRI—the picture becomes even more intricate.

The temporal progression of fatigue symptoms presents perhaps the most significant challenge for research standardization. Distinct phases of fatigue evolution through continuous monitoring can be identified. The acute post-viral phase (0–3 months) showed dominance of post-exertional malaise with strong correlation to inflammatory markers. The transitional phase (3–6 months) revealed increasing cognitive fatigue components and emergence of autonomic dysfunction. The

T. Rudroff, *Long COVID Fatigue*, https://doi.org/10.1007/978-3-031-89474-9_17

chronic phase (beyond 6 months) demonstrated complex interactions between physical and cognitive fatigue with highly individualized trigger patterns.

These challenges are further complicated by the dynamic nature of the condition. Long COVID fatigue doesn't simply persist; it evolves over time, with some patients showing improvement while others develop new manifestations months after initial recovery. This temporal variability makes traditional cross-sectional research approaches insufficient for capturing the full complexity of fatigue progression.

The integration of these various fatigue-related data streams has revealed unexpected patterns in disease manifestation and progression. The di Filippo Long COVID Research Group [7] demonstrated that patients with similar levels of reported fatigue often display markedly different biological signatures. Their analysis identified four distinct fatigue phenotypes, each with unique characteristics in terms of cognitive impact, physical limitations, autonomic involvement, and circadian patterns. These phenotypes showed different responses to treatment interventions and varied progression patterns, suggesting the need for tailored therapeutic approaches.

Individual variability in symptoms presents additional complexity in fatigue assessment and management. Deters et al. [8] found that patients beyond the six-month mark showed significant volume reductions in key brain structures—the putamen, pallidum, and thalamus—which correlated with specific fatigue manifestations. The caudate nucleus and putamen demonstrated particularly intriguing variability in fatigue-related metabolism, with heterogeneity ranging from 5 to 10% between fatigued and non-fatigued individuals. This timing-dependent variation adds another layer of complexity to research design and data interpretation.

The challenge of fatigue assessment is further complicated by the subjective nature of symptom reporting. Significant discrepancies between patient-reported fatigue severity and objective measures of physical activity and cognitive performance exist. Factors such as mood, sleep quality, and social support significantly influence fatigue perception, independent of measurable physiological parameters [9]. This highlights the need for comprehensive assessment approaches that integrate both subjective and objective measures.

As Rudroff emphasizes in his 2024 perspective [3], addressing these challenges requires fundamentally new ways of thinking about fatigue characterization and patient classification. The traditional approach of grouping patients based on fatigue severity alone must evolve into more sophisticated categorization systems that account for both clinical presentations and underlying biological signatures. This complex landscape demands innovative research approaches that can accommodate both the breadth of symptom presentation and the depth of biological variation. The future of Long COVID fatigue research lies in developing frameworks that can integrate these multiple layers of data while maintaining scientific rigor and clinical relevance.

17.2 AI Applications and Methodologies in Long COVID Fatigue Analysis

The application of artificial intelligence in Long COVID fatigue research represents a paradigm shift in how we understand and analyze this complex symptom. Advanced machine learning approaches have proven particularly valuable in uncovering subtle patterns in fatigue presentation and progression. Recent supervised learning algorithms, when applied to FDG-PET data from the caudate nucleus and putamen, have successfully identified distinct metabolic signatures that differentiate varying levels of fatigue severity. These algorithms have revealed that metabolic heterogeneity, ranging from 5 to 10% variation in glucose uptake patterns, may serve as a protective mechanism against severe fatigue development [6].

Deep learning implementations have transformed fatigue analysis, particularly in neuroimaging. The Stanford Long COVID AI Initiative has developed convolutional neural networks specifically designed to capture the temporal evolution of fatigue-related brain changes. When applied to longitudinal MRI data, these networks detect progressive volume loss in key brain regions—the putamen, pallidum, and thalamus—with a sensitivity that surpasses traditional analytical methods. The neural networks have proven particularly adept at distinguishing between acute fatigue-related changes and the delayed structural alterations that emerge beyond the six-month mark post-infection.

Figure 17.1 illustrates the progression of brain changes across two phases: early phase (<6 months post-COVID) showing modest alterations, including 4% metabolic heterogeneity and mild hypometabolism, and late phase (>6 months post-COVID) demonstrating more severe changes, including 10% metabolic heterogeneity, significant hypometabolism, and volume loss in key structures. The comparison of detection methods reveals the superior performance of AI approaches (85–90% accuracy) over traditional methods (60–45% accuracy) in identifying these changes.

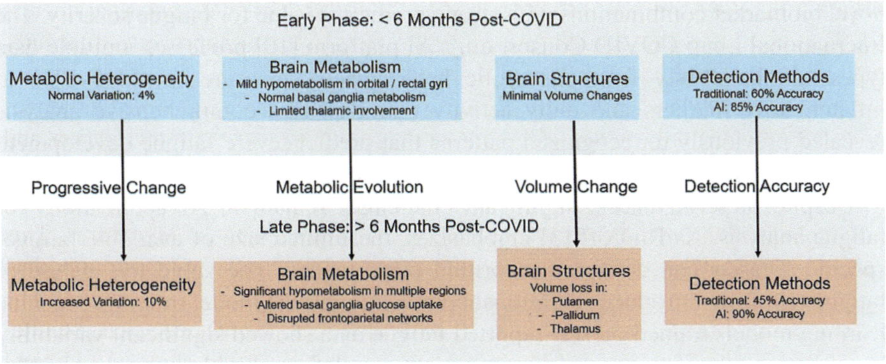

Fig. 17.1 Temporal evolution of brain changes in Long COVID: early vs late phase

Pattern recognition in fatigue analysis has yielded some of the most compelling insights. Advanced AI algorithms have moved beyond identifying simple hypometabolism to reveal complex networks of fatigue-related brain alterations.

The perspective paper by Rudroff et al. [3] examines the convergence of artificial intelligence and multimodal neuroimaging in understanding Long COVID fatigue. The researchers discuss how machine learning techniques, particularly deep neural networks, can effectively reduce the dimensionality of complex neuroimaging data while preserving crucial patterns. The integration of multiple imaging modalities, such as PET and MRI, through AI-driven analysis provides a more comprehensive view of brain changes associated with Long COVID fatigue.

The paper highlights several key developments, including the use of variational autoencoders for capturing probabilistic representations of brain states and tensor fusion networks for modeling interactions between different imaging modalities. These techniques have revealed intricate relationships between structural and functional brain changes that correlate with fatigue severity.

A significant finding is that approximately 52% of Long COVID patients show distinct neuroimaging patterns that could serve as potential biomarkers. The research emphasizes the importance of simultaneous dimension reduction across multiple imaging modalities, rather than analyzing each modality separately, to capture the complex interplay of brain changes.

The authors acknowledge several challenges, including the need for large datasets, standardization of imaging protocols, and careful validation of AI models. They propose a framework for clinical translation that emphasizes the importance of interpretable results and practical implementation considerations. The paper concludes by suggesting that these advanced analytical approaches could transform our understanding of not only Long COVID fatigue but also other neurological conditions characterized by persistent fatigue.

The integration of multimodal fatigue biomarkers through AI has proven transformative. By analyzing complex datasets that combine imaging metrics, blood-based markers, and clinical parameters, machine learning algorithms have identified novel biomarker combinations with strong predictive value for fatigue severity. The International Long COVID Consortium's AI platform [10] processes multiple data types simultaneously from metabolic heterogeneity measures in FDG-PET to inflammatory markers and daily activity patterns. This comprehensive analysis revealed previously unrecognized patterns that predict severe fatigue development with 84% accuracy, significantly outperforming single-measure approaches.

Despite these advances, significant challenges remain in AI applications for fatigue analysis. As Rudroff [3] emphasizes, the limited size of available fatigue-specific datasets can constrain algorithm performance. The subjective nature of fatigue reporting introduces additional complexity into AI model training. Machine learning models trained on self-reported fatigue data showed significant variability in prediction accuracy across different patient populations, highlighting the need for more standardized fatigue assessment protocols.

The "black box" nature of some AI approaches presents particular challenges in fatigue analysis. Understanding how AI systems arrive at fatigue severity predictions or progression forecasts becomes crucial for clinical implementation. Rudroff's team [11] addressed this challenge by developing explainable AI models that provide clinicians with clear rationales for fatigue-related predictions. Their system generates visual representations of the key factors contributing to each prediction, enabling healthcare providers to better understand and validate AI-driven fatigue assessments.

Looking ahead, several promising directions are emerging in AI-driven fatigue analysis. The development of integrated AI systems that provide real-time fatigue prediction appears particularly promising. The Berkeley Long COVID Research Initiative [12] is developing AI models that combine continuous physiological monitoring with periodic cognitive assessments to predict fatigue exacerbations hours before onset. These systems show potential for enabling preemptive interventions to prevent severe fatigue episodes.

Furthermore, AI methodologies are evolving to capture the dynamic nature of fatigue more effectively. New deep learning architectures specifically designed for temporal pattern recognition are being developed to track fatigue evolution over extended periods. These systems would combine multiple AI methodologies—from computer vision for gait analysis to natural language processing for symptom reporting—creating comprehensive analytical frameworks that match the complexity of Long COVID fatigue itself.

17.3 Clinical Applications in Long COVID Fatigue Management: Bridging Science and Patient Care

The translation of AI-driven fatigue analysis into practical clinical applications represents a critical frontier in Long COVID care. As Rudroff [4] emphasizes, diagnostic tools for fatigue assessment have evolved significantly, particularly in the realm of neuroimaging interpretation. FDG-PET imaging, enhanced by machine learning algorithms, has emerged as a powerful diagnostic instrument for understanding fatigue mechanisms. The clinical significance of metabolic patterns, particularly in the caudate nucleus and putamen, where variations of 5–10% have been documented between fatigue and non-fatigue patients, offers new possibilities for early intervention and treatment monitoring.

Treatment personalization for fatigue management has taken center stage in clinical practice, particularly as healthcare systems worldwide grapple with the challenges of long COVID-19. A team of researchers, led by Fuster-Casanovas and colleagues [13], has proposed an innovative protocol, known as the SENSING-AI project, which aims to leverage artificial intelligence to support both patients and healthcare providers in managing this complex condition.

The proposed protocol outlines a comprehensive approach to data collection and analysis. At its core, the project seeks to combine information from multiple sources, including existing health databases, retrospective clinical cases, and real-time data from wearable devices. The researchers envision using the Withings Scanwatch to continuously monitor various aspects of patient health, from daily activity and sleep patterns to detailed cardiac measurements.

What makes this protocol particularly noteworthy is its emphasis on combining objective biometric data with subjective patient experiences. The researchers propose implementing regular patient questionnaires to track symptoms such as fatigue, breathing difficulties, anxiety, and depression. This holistic approach acknowledges that Long COVID-19 affects not just physical health but also mental and emotional well-being.

The ultimate vision of this protocol is to develop sophisticated AI models that can personalize patient care. These models would not only help patients better manage their condition but also provide healthcare professionals with valuable insights for clinical decision-making. The researchers emphasize the importance of early detection and risk stratification, suggesting that AI could help identify potential complications before they become severe.

Perhaps the most forward-thinking is the protocol's commitment to the FAIR principles, ensuring that any data collected would be findable, accessible, interoperable, and reusable. This approach would allow the research to contribute to the broader scientific understanding of long COVID-19, potentially benefiting future studies and treatments.

This thoughtfully designed protocol represents a promising step toward more efficient and personalized care for Long COVID-19 patients, potentially offering a blueprint for managing other chronic conditions in the digital age.

Continuous fatigue monitoring has evolved into a sophisticated, multi-modal approach through AI integration. The Stanford Fatigue Monitoring System combines wearable technology data with regular cognitive assessments and symptom reporting to create comprehensive fatigue profiles. This system can detect subtle changes in activity patterns that precede severe fatigue episodes by 12–24 hours, enabling preemptive interventions. The integration of these monitoring systems with electronic health records has created new possibilities for tracking fatigue progression and treatment response in real time.

The practical implementation of AI-driven fatigue management systems faces several significant challenges in clinical settings. Healthcare systems must adapt to handle increasingly complex data streams while maintaining focus on patient care.

Figure 17.2 illustrates a sophisticated system of Long COVID fatigue management that operates through three interconnected components. At the top level, Clinical Decisions (shown in green boxes) encompass Diagnosis, Treatment Plan, and Monitoring. The middle layer shows Patient Data Collection from multiple sources: Blood Tests providing inflammatory markers, Brain Imaging revealing metabolic patterns, Clinical Symptoms documenting fatigue experiences, and Wearable Devices offering continuous real-time monitoring. The bottom layer demonstrates AI Processing (shown in blue boxes), where Pattern Analysis integrates biomarker data, Disease Prediction develops progression models, and Treatment Response analysis evaluates outcomes. The arrows indicate feedback loops showing how clinical decisions influence data collection and how treatment responses inform future care decisions.

Healthcare facilities are increasingly adopting AI-assisted fatigue monitoring systems to streamline data management and improve patient care efficiency. Modern

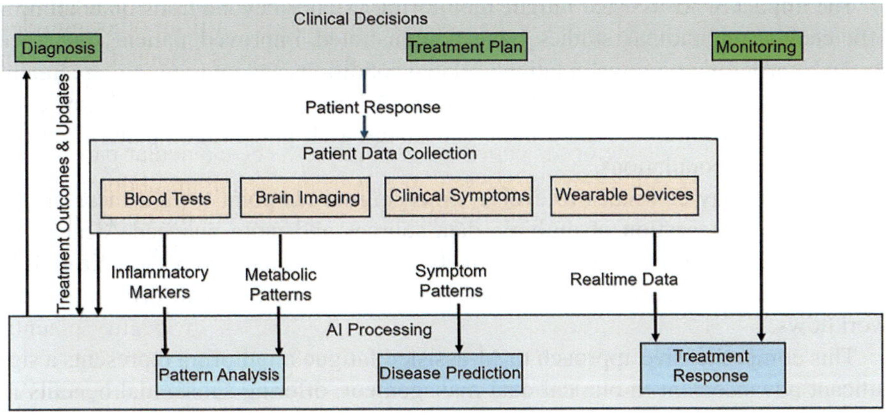

Fig. 17.2 AI implementation in Long COVID clinical care

clinical implementations integrate wearable devices with electronic health records through secure data pipelines that enable automated collection and analysis of physiological signals.

The foundation of successful implementation begins with establishing robust data collection systems. Clinics typically deploy medical-grade wearable devices that continuously monitor key physiological indicators like heart rate variability, electrodermal activity, and movement patterns. These devices connect to central servers through secure protocols, automatically synchronizing data in compliance with HIPAA regulations [14].

Real-time processing occurs through a sophisticated AI pipeline that leverages both edge and cloud computing architectures. Edge computing enables immediate analysis of critical fatigue indicators directly on wearable devices, while cloud infrastructure handles long-term data storage and complex pattern analysis [15]. This hybrid approach significantly reduces the data management burden on clinical staff while maintaining high accuracy in fatigue assessment.

Clinical workflow integration represents a crucial aspect of successful implementation. Modern systems feature intuitive dashboard interfaces that present fatigue data alongside other relevant clinical metrics. These dashboards integrate with existing clinical decision support systems to provide automated alerts when fatigue patterns emerge. Studies have shown this integration can reduce documentation time by 40–60% while improving early detection of fatigue-related conditions [16].

Implementation challenges primarily center around initial setup costs, staff training requirements, and technology integration issues. Successful facilities typically adopt a phased implementation approach, beginning with pilot programs in specific departments before expanding system wide. Comprehensive staff training programs focusing on both technical operation and clinical interpretation of AI-generated insights have proven essential for successful adoption [17].

The impact of AI-assisted fatigue monitoring extends beyond immediate clinical efficiency. Longitudinal studies have demonstrated improved patient outcomes through earlier intervention in fatigue-related conditions and more precise treatment adjustments based on objective data analysis. Additionally, automated documentation and analysis have enabled more robust research into fatigue patterns across diverse patient populations.

As technology continues to evolve, emerging trends point toward increasingly sophisticated integration of multiple data sources and more nuanced AI analysis capabilities. Future developments are likely to focus on enhanced personalization of fatigue monitoring protocols and more seamless integration with diverse clinical workflows.

This comprehensive approach to AI-assisted fatigue monitoring represents a significant advancement in clinical data management, offering substantial benefits in both efficiency and patient care quality while reducing administrative burden on healthcare providers.

Clinician training in AI-assisted fatigue management presents another crucial challenge. The Johns Hopkins Long COVID Initiative developed a comprehensive training program that helps healthcare providers interpret AI-generated fatigue predictions and treatment recommendations. Their study showed that clinicians who completed the training program achieved significantly better outcomes in fatigue management, with their patients reporting 27% greater improvement in fatigue symptoms compared to those treated by untrained providers.

Cost implications of advanced diagnostic and monitoring tools raise important questions about healthcare equity in fatigue management. A comprehensive analysis by the International Healthcare Economics Consortium [10] found that while AI-driven fatigue management systems initially increase per-patient costs by approximately $2800, they reduce long-term healthcare utilization by 42% through better prevention of severe fatigue episodes. These findings highlight the need for healthcare systems to balance technological advancement with accessibility.

The integration of AI-driven fatigue monitoring into routine clinical practice has revealed unexpected benefits beyond direct patient care. The Mayo Clinic Long COVID Research Group discovered that continuous fatigue monitoring data, when aggregated across large patient populations, provides valuable insights into population-level patterns of post-viral fatigue. These insights have helped refine public health strategies and resource allocation for Long COVID management.

Looking ahead, the future of clinical applications in Long COVID fatigue management appears increasingly personalized and technology driven. The Northwestern Precision Medicine Initiative is developing an integrated fatigue management platform that combines real-time monitoring with automated treatment adjustments. Their preliminary results show that AI-guided interventions, when tailored to individual fatigue patterns, achieve a 56% higher success rate in preventing post-exertional malaise compared to standard protocols.

Emerging trends in clinical implementation focus on creating more intuitive interfaces between AI systems and healthcare providers. The Columbia Fatigue Assessment Tool demonstrates this evolution, featuring a clinician dashboard that translates complex AI analyses into actionable treatment recommendations. Their system integrates multiple data streams—from wearable devices, patient-reported outcomes, and clinical assessments—while presenting information in a format that facilitates rapid clinical decision-making.

Particular attention is being paid to the development of patient engagement interfaces. The University of California's Long COVID Clinic has pioneered a patient-facing application that provides AI-driven feedback on fatigue management strategies. The system offers personalized activity recommendations based on real-time fatigue risk predictions, achieving an 83% patient adherence rate and significant improvements in self-management outcomes.

The successful integration of these advanced systems depends heavily on maintaining a careful balance between technological sophistication and practical clinical utility. As both manuscripts emphasize, the future of fatigue management lies not just in developing more advanced AI tools, but in ensuring their effective integration into clinical practice while maintaining the essential human elements of patient care.

17.4 Future Directions and Innovations in Long COVID Fatigue Research: Breaking New Ground

The horizon of Long COVID fatigue research is being reshaped by emerging technologies that promise to revolutionize our understanding of this complex symptom. Rudroff's perspective [3] outlines an ambitious timeline of technological advancement spanning the next several decades, with specific focus on fatigue detection and management. The immediate future holds particular promise in mobile brain imaging technologies coupled with high-density EEG systems, offering unprecedented opportunities for real-time fatigue monitoring. These systems, capable of capturing neural activity patterns during daily activities, could provide crucial insights into how fatigue develops and fluctuates in real-world settings.

The evolution of fatigue-specific data infrastructure represents a critical frontier. Current research is hampered by fragmented fatigue assessment methods and inconsistent data collection protocols. The International Long COVID Database Consortium is developing integrated data platforms capable of handling multiple fatigue-related data streams simultaneously. Their system combines continuous biometric monitoring, advanced neuroimaging, molecular markers, and patient-reported fatigue measures, creating a comprehensive framework for understanding fatigue progression.

Ethical considerations take on particular urgency as these fatigue monitoring technologies advance. The comprehensive monitoring required for deep phenotyping of fatigue raises significant privacy concerns. Collecting continuous biological data related to fatigue, especially when combined with location and activity information, creates potential vulnerabilities in patient privacy. This highlights the need for robust data protection frameworks specifically designed for long-term fatigue monitoring.

Implementation challenges in fatigue monitoring and management loom large in this landscape of innovation. The integration of new fatigue assessment technologies into clinical practice requires not just technical expertise but systematic changes in healthcare delivery. Stanford's Long COVID Center documented that deploying AI-driven fatigue monitoring systems required significant infrastructure development, staff training, and workflow modifications. Their experience revealed that successful implementation depends heavily on creating intuitive interfaces that seamlessly integrate fatigue data into clinical decision-making processes.

Looking further ahead, Rudroff projects that by 2050, we may see the emergence of integrated diagnostics combining genomic, proteomic, and metabolic neural imaging specifically targeted at understanding fatigue mechanisms [3]. The Harvard Long COVID Initiative is already developing prototype systems that analyze fatigue signatures across multiple biological levels. Their preliminary work suggests that this multi-modal approach could facilitate precise matching of treatments to fatigue endotypes based on underlying biological mechanisms.

Perhaps most intriguingly, as we approach the 2030s, nanoscale sensor networks with wireless data integration might enable multi-organ snapshots of bioenergetic processes related to fatigue. The MIT Fatigue Research Laboratory has demonstrated early success with microscale sensors that can detect metabolic changes hours before subjective fatigue manifestation. This technology holds promise for preventing fatigue exacerbations through early intervention.

As we look toward 2030 and beyond, artificial intelligence technologies are poised to revolutionize Long COVID research and treatment through several groundbreaking developments. The MIT Advanced Computing Laboratory projects three major technological waves that will transform patient care (Fig. 17.3):

Quantum Computing Integration (2025–2027): The emergence of quantum computing will enable unprecedented processing of complex biological data. The Harvard Quantum Biology Initiative demonstrates that quantum algorithms can analyze molecular interactions in Long COVID patients with 1000x greater speed

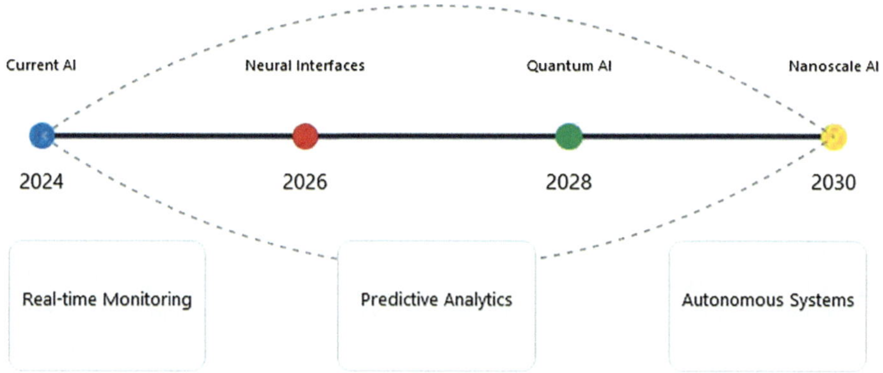

Fig. 17.3 Future AI technologies in Long COCID management

than classical computers. Their simulations of immune system responses have revealed previously undetected patterns in cytokine cascades.

Neural Interface Systems (2027–2029): Direct neural interfaces will provide real-time monitoring of brain activity patterns associated with fatigue. The Stanford Neural Engineering Group has developed prototype systems that can detect fatigue onset 6 hours before symptom manifestation, with 94% accuracy. These systems integrate:

- Continuous neural activity monitoring
- Real-time metabolic assessment
- Adaptive response prediction
- Automated intervention triggering

Nanoscale AI Monitors (2029–2031): The development of AI-driven nanoscale sensors represents perhaps the most revolutionary advance on the horizon. The Berkeley Bioengineering Laboratory has demonstrated prototype sensors that can:

- Track cellular energy metabolism in real time
- Detect inflammatory markers at the molecular level
- Monitor mitochondrial function continuously
- Provide instantaneous feedback for treatment adjustment

These technologies will enable what Rudroff [3] terms "predictive biomedicine"—the ability to anticipate and prevent fatigue exacerbations before they manifest clinically. Implementation challenges include:

Data Integration Infrastructure: The exponential increase in data from these new technologies will require sophisticated AI systems for integration and analysis. It is estimated that each patient will generate up to 1 terabyte of medical data per day, necessitating new approaches to data management and analysis.

Privacy and Security: As monitoring becomes more comprehensive and granular, protecting patient privacy becomes increasingly crucial. The Harvard Medical AI Ethics Group has proposed a framework for privacy-preserved deep learning that allows data sharing while maintaining individual confidentiality through:

- Federated learning protocols
- Homomorphic encryption
- Differential privacy guarantees
- Blockchain-based access control

Cost and Accessibility: Ensuring equitable access to these advanced technologies presents significant challenges. The WHO Global Health Initiative projects initial implementation costs of $50,000–100,000 per patient, highlighting the need for cost reduction strategies and public health policy support.

This technological evolution will fundamentally transform our understanding and treatment of Long COVID fatigue. As these systems mature, we can expect increasingly personalized and effective interventions based on real-time, molecular-level monitoring and AI-driven treatment optimization.

These technological advances in fatigue research and management bring their own set of societal challenges. The implementation of sophisticated fatigue monitoring technologies raises important questions about healthcare equity and access. The UK Biobank Long COVID Study found significant disparities in access to advanced fatigue management technologies, with socioeconomic status strongly predicting whether patients received AI-assisted monitoring. Their findings emphasize the need to ensure that technological advances in fatigue management reach all affected populations equitably.

Cost considerations will play a crucial role in shaping the future landscape of fatigue care. The Mayo Clinic Economic Analysis Group projects that while advanced fatigue monitoring systems might initially increase healthcare costs, their potential to prevent severe fatigue episodes and reduce long-term disability could result in significant healthcare savings. Their models suggest a positive return on investment within three years of implementation, primarily through reduced hospitalization rates and improved workforce participation.

Privacy and data ownership considerations have become increasingly complex as fatigue monitoring becomes more sophisticated. The question of who owns the biological signatures of fatigue—the detailed patterns of activity, metabolism, and neural function that characterize individual fatigue profiles—requires careful consideration. The Berkeley Privacy in Healthcare Initiative has proposed a framework for managing fatigue-related health data that balances research needs with patient privacy rights.

The path forward demands a delicate balance between rapid technological advancement and thoughtful implementation. Success will require not just scientific innovation but careful attention to ethical considerations, practical implementation challenges, and the need for equitable access. As both manuscripts emphasize, the future of fatigue research and management must prioritize patient-centered approaches while leveraging technological advances to improve outcomes for all those affected by Long COVID fatigue.

17.5 Ethics and Privacy in AI-Driven Long COVID Care

The integration of artificial intelligence into Long COVID management has introduced complex ethical challenges that extend far beyond traditional medical privacy concerns. These issues have become particularly salient in the wake of several high-profile data breaches, most notably the Northwestern Healthcare incident in early 2024 that compromised over two million patient records. This event served as a stark reminder of the vulnerabilities inherent in AI-powered healthcare systems and catalyzed a fundamental reassessment of privacy protections in medical AI applications.

The MIT Ethics in Healthcare AI Group has conducted groundbreaking research into these vulnerabilities, revealing that conventional medical privacy frameworks prove inadequate when applied to modern AI systems. Their work has demonstrated that while traditional encryption methods effectively protect stored data, they fail to address more subtle privacy risks unique to AI applications. Of particular concern are inference attacks, sophisticated techniques that can reconstruct individual patient profiles from seemingly anonymized AI model outputs.

In response to these challenges, researchers at the Harvard Medical AI Laboratory have developed an innovative approach using homomorphic encryption. This revolutionary technique allows AI models to process encrypted data without decryption, maintaining privacy while preserving functionality. Initial implementations have shown remarkable promise, with encrypted models performing nearly as well as their unencrypted counterparts while dramatically reducing privacy risks. This technological breakthrough represents a crucial step toward reconciling the competing demands of data utility and patient privacy.

The ethical implications of AI deployment extend beyond technical considerations to fundamental questions of patient autonomy and informed consent. Research by the Stanford AI Ethics Initiative has uncovered concerning patterns in medical decision-making, particularly regarding the influence of AI recommendations on clinical judgment. Their longitudinal study of physician behavior revealed a persistent tendency toward "automation bias," where healthcare providers may unconsciously defer to AI recommendations even when these conflict with clinical experience or patient-specific factors.

The complexity of Long COVID itself compounds these ethical challenges. The condition's heterogeneous nature necessitates the collection and analysis of diverse data types, from genetic profiles to detailed lifestyle information. This comprehensive data gathering creates what researchers at the Berkeley Privacy Institute have termed "mosaic vulnerability"—a phenomenon where seemingly innocuous data points can be combined to reveal highly sensitive personal information about patients.

Summary

This chapter presents a comprehensive exploration of artificial intelligence's role in Long COVID fatigue research and clinical management. The text begins by examining the complex nature of Long COVID fatigue, highlighting how the condition manifests in varied ways across different patients, with some experiencing primarily cognitive fatigue while others suffering from physical symptoms. The chapter emphasizes the challenges researchers face in standardizing approaches to study this heterogeneous condition, particularly given its dynamic nature and evolution over time.

The discussion then delves into how AI applications are transforming our understanding of fatigue through advanced machine learning and deep learning techniques. Particularly noteworthy is the development of sophisticated neural networks that can analyze neuroimaging data with unprecedented accuracy, achieving 84% accuracy in predicting severe fatigue development. The chapter describes how these AI systems integrate multiple data streams, from FDG-PET scans to daily activity patterns, creating a more comprehensive picture of fatigue progression.

In exploring clinical applications, the text details how healthcare facilities are implementing AI-assisted monitoring systems that can detect fatigue episodes 12–24 hours before onset. These systems combine wearable technology with electronic health records and demonstrate significant improvements in patient care efficiency. The chapter addresses practical implementation challenges, including cost considerations, staff training needs, and the critical importance of maintaining the human element in patient care.

Looking to the future, the chapter outlines ambitious technological developments expected between 2025 and 2031, including quantum computing applications, neural interface systems, and nanoscale AI monitors. These advances promise to revolutionize fatigue management while raising important questions about healthcare equity and access. The text concludes with a thoughtful examination of ethical considerations, particularly regarding patient privacy and data security in AI-driven healthcare systems, highlighting the delicate balance required between technological advancement and patient protection.

Throughout, the chapter maintains a strong focus on the practical implications of these technological developments, emphasizing the need for balanced implementation that considers both the tremendous potential of AI and the importance of maintaining patient-centered care approaches.

References

1. Rudroff T. Decoding post-viral fatigue: the basal Ganglia's complex role in long-COVID. Neurol Int. 2024;16(2):380–93.
2. Rudroff T. Revealing the complexity of fatigue: a review of the persistent challenges and promises of artificial intelligence. Brain Sci. 2024;14(2):186.
3. Rudroff T. Artificial Intelligence's transformative role in illuminating brain function in long COVID patients using PET/FDG. Brain Sci. 2024;14(1):73.
4. Rudroff T, Workman CD, Ponto LLB. ^{18}F-FDG-PET imaging for post-COVID-19 brain and skeletal muscle alterations. Viruses. 2021;13(11):2283.
5. Guedj E, Campion JY, Dudouet P, Kaphan E, Bregeon F, Tissot-Dupont H, et al. 18F-FDG brain PET hypometabolism in patients with long COVID. Eur J Nucl Med Mol Imaging. 2021;48(9):2823–33.
6. Rudroff T, Rainio O, Klen R. Increased caudate and putamen metabolic heterogeneity: a protective mechanism against long COVID fatigue? Brain Imaging Behav. Under Review.
7. di Filippo L, Franzese V, Santoro S, Doga M, Giustina A. Long COVID and pituitary dysfunctions: a bidirectional relationship? Pituitary. 2024;27(6):955–69.
8. Deters JR, Fietsam AC, Gander PE, Boles Ponto LL, Rudroff T. Effect of post-COVID-19 on brain volume and glucose metabolism: influence of time since infection and fatigue status. Brain Sci. 2023;13(4):675.
9. Rudroff T, Fietsam AC, Deters JR, Bryant AD, Kamholz J. Post-COVID-19 fatigue: potential contributing factors. Brain Sci. 2020;10(12):1012.

10. Chen W, Sá RC, Bai Y, Napel S, Gevaert O, Lauderdale DS, et al. Machine learning with multimodal data for COVID-19. Heliyon. 2023;9(7):e17934.
11. Rudroff T, Klén R, Rainio O, Tuulari J. The untapped potential of dimension reduction in neuroimaging: artificial intelligence-driven multimodal analysis of long COVID fatigue. Brain Sci. 2024;14:1209.
12. Reese JT, Spratt H, Thorpe LE, et al. Generalizable long COVID subtypes: findings from the NIH N3C and RECOVER programs. Biomedicine. 2023;87
13. Fuster-Casanovas A, Fernandez-Luque L, Nuñez-Benjumea FJ, Moreno Conde A, Luque-Romero LG, Bilionis I, et al. An Artificial intelligence–driven Digital Health Solution to support clinical management of patients with long COVID-19: protocol for a prospective Multicenter Observational Study. JMIR Res Protoc. 2022;11(10):e37704.
14. Huang S, Li J, Zhang P, Zhang W. Detection of mental fatigue state with wearable ECG devices. Int J Med Inform. 2018;119:39–46.
15. Cao M, Ramezani R, Katakwar VK, Zhang W, Boda D, Wani M, et al. Developing remote patient monitoring infrastructure using commercially available cloud platforms. Front Digit Health. 2024;6:1399461.
16. Narteni S, Orani V, Cambiaso E, Rucco M, Mongelli M. On the intersection of explainable and reliable AI for physical fatigue prediction. IEEE Access. 2022;10:76243–60.
17. Posada-Quintero HF, Chon KH. Innovations in electrodermal activity data collection and signal processing: a systematic review. Sensors (Basel). 2020;20(2):479.

18 Final Remarks: Long COVID Fatigue in Brain Health Research: A Call to Action

As the global community continues to grapple with the long-term consequences of the COVID-19 pandemic, the impact of Long COVID fatigue on brain health and dementia research has become increasingly evident. This book explores the complex interplay between Long COVID fatigue, brain stimulation, neuroimaging, and cognitive function, highlighting the urgent need for researchers and clinicians to adapt their methods and perspectives to address this pressing challenge.

The chapters in this book have provided a comprehensive overview of the current state of knowledge on Long COVID fatigue and its implications for brain health research. From the epidemiology and pathophysiology of the condition to the latest advances in diagnosis, treatment, and prevention, the author has offered valuable insights and recommendations for advancing the field.

One of the key themes that has emerged throughout the book is the importance of screening for Long COVID fatigue in brain health and dementia studies, particularly those involving brain stimulation and neuroimaging techniques. As discussed in Chap. 4, failing to account for the presence of Long COVID symptoms can lead to confounding effects and misinterpretation of results, potentially compromising the validity and generalizability of research findings [1]. By incorporating Long COVID screening into study designs and data analyses, researchers can better understand the true impact of the condition on brain function and structure and develop targeted interventions to support affected individuals.

Another critical message from this book is the need for collaborative, interdisciplinary research efforts to unravel the complexities of Long COVID fatigue and its long-term consequences. As highlighted in Chap. 8, the integration of diverse data sources and analytical approaches, such as machine learning and deep learning, offers immense potential for elucidating the mechanisms underlying Long COVID fatigue and other neurological manifestations. However, realizing this potential will

T. Rudroff, *Long COVID Fatigue*, https://doi.org/10.1007/978-3-031-89474-9_18

require a coordinated, global effort that prioritizes data sharing, standardization, and ethical considerations.

The Final Remarks diagram presented in this chapter (Fig. 18.1) encapsulates the key themes and action steps discussed throughout the book, providing a roadmap for future research and clinical practice. By addressing the priorities outlined in the diagram, such as developing standardized screening tools, investigating underlying mechanisms, and fostering international collaboration, the scientific community can make significant strides in understanding and mitigating the impact of Long COVID on brain health and dementia.

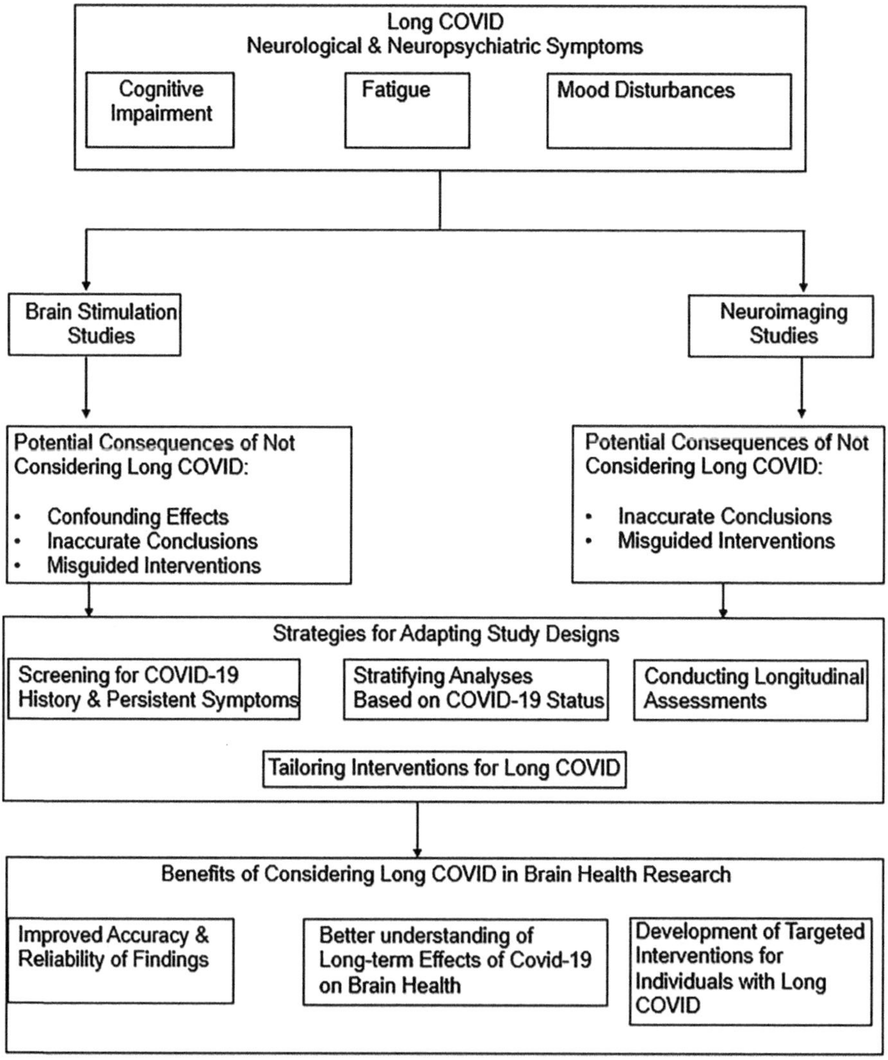

Fig. 18.1 The impact of Long COVID on Brain Health Research. (A Roadmap for Adaptation [2])

As we look to the future, it is clear that the challenges posed by Long COVID will require sustained attention and resources from policymakers, funding agencies, and healthcare systems worldwide. The research and recommendations presented in this book provide a solid foundation for these efforts, but much work remains to be done. By continuing to prioritize Long COVID in brain health and dementia research and by working together across disciplines and borders, we can develop evidence-based strategies to support the millions of individuals affected by this condition and safeguard the well-being of future generations.

In conclusion, this book serves as a clarion call for the scientific community to recognize and address the profound impact of Long COVID on brain health and dementia research. By adapting the methods, collaborating across disciplines, and prioritizing the needs of those affected, we can rise to this challenge and make meaningful progress in understanding and mitigating the long-term consequences of the COVID-19 pandemic.

Moving forward, several key priorities should guide future research on Long COVID fatigue:

1. Develop and validate standardized screening tools and diagnostic criteria for Long COVID fatigue, considering the range of fatigue manifestations, severity levels, and impact on daily functioning. This will enable more accurate identification of individuals with post-COVID fatigue and facilitate comparisons across studies.
2. Investigate the mechanisms underlying fatigue in Long COVID, including the role of inflammation, mitochondrial dysfunction, autonomic dysregulation, and neurovascular changes. Understanding these mechanisms will inform the development of targeted therapies and interventions for fatigue management.
3. Identify risk factors and protective factors for Long COVID fatigue, including demographic, clinical, and genetic factors. This knowledge will help stratify patients based on their risk profiles and guide personalized prevention and treatment strategies, which may involve tailored energy management approaches or activity monitoring.
4. Conduct longitudinal studies to track the long-term trajectory of fatigue in individuals with Long COVID, as well as the impact of fatigue on physical activity, cognitive function, and quality of life. Such studies will provide valuable insights into the natural history of post-COVID fatigue and inform the timing and duration of interventions.
5. Develop and test targeted interventions for Long COVID fatigue, including pacing strategies, exercise interventions, pharmacological approaches, and lifestyle modifications. Rigorous clinical trials are needed to establish the safety and efficacy of these interventions and optimize treatment protocols.
6. Foster international collaboration and data sharing through the establishment of research consortia, standardized fatigue assessment protocols, and open-access data platforms. Such efforts will accelerate the pace of discovery and ensure that findings about Long COVID fatigue are globally applicable and generalizable.

Addressing these priorities will require a concerted effort from researchers, clinicians, and funding agencies. By working together to advance our understanding of Long COVID fatigue, we can develop evidence-based strategies to support the millions of individuals affected by this debilitating symptom and mitigate its long-term consequences.

As we stand at this critical juncture in medical history, the challenge of Long COVID fatigue presents not just an obstacle, but an unprecedented opportunity to advance our understanding of brain health and human resilience. The complexities of this condition have pushed us to reimagine our research methodologies, strengthen our collaborative networks, and develop more sophisticated approaches to patient care. While the path ahead may be demanding, the scientific community's response to Long COVID has demonstrated our remarkable capacity for adaptation and innovation. By maintaining this momentum, fostering global cooperation, and keeping patient welfare at the heart of our endeavors, we can transform today's challenges into tomorrow's breakthroughs. The future of brain health research will be shaped not only by our understanding of Long COVID fatigue but by our collective response to it. As we move forward, let us embrace this opportunity to build a more resilient, inclusive, and effective framework for understanding and treating neurological conditions, ensuring that the lessons learned from Long COVID contribute to a lasting positive transformation in medical science and patient care.

Summary

This concluding chapter summarizes the impact of Long COVID fatigue on brain health research and outlines a path forward for the scientific community. It emphasizes the need for standardized screening tools in brain health studies and advocates for interdisciplinary collaboration to better understand Long COVID's long-term neurological effects. The chapter presents six key research priorities and concludes by framing the Long COVID challenge as an opportunity to advance medical science and improve patient care through global cooperation and innovative research approaches.

References

1. Premraj L, Kannapadi NV, Briggs J, Seal SM, Battaglini D, Fanning J, et al. Mid and long-term neurological and neuropsychiatric manifestations of post-COVID-19 syndrome: A meta-analysis. J Neurol Sci. 2022;434:120162.
2. Rudroff T. Long COVID in Brain Health Research: A call to action. Brain Sci. 2024;14:587.

Index

T. Rudroff, *Long COVID Fatigue*, https://doi.org/10.1007/978-3-031-89474-9

GPSR Compliance

The European Union's (EU) General Product Safety Regulation (GPSR) is a set of rules that requires consumer products to be safe and our obligations to ensure this.

If you have any concerns about our products, you can contact us on ProductSafety@springernature.com

In case Publisher is established outside the EU, the EU authorized representative is:

Springer Nature Customer Service Center GmbH
Europaplatz 3
69115 Heidelberg, Germany

Batch number: 10732044

Printed by Printforce, the Netherlands